Miner Wisdom

By

Stephen A. Wzorek

ISBN: 1-4107-9306-0 (e-book)
ISBN: 1-4107-9307-9 (Paperback)
ISBN: 1-4140-4486-0 (Dust Jacket)

Library of Congress Control Number: 2003098144

This book is printed on acid free paper.

Printed in the United States of America
Bloomington, IN

1stBooks – rev. 12/23/03

Synopsis

Miner Wisdom depicts the many ways that greed harms innocent people and strips land of its natural beauty.

The story describes the hardships endured by a Polish immigrant family as they leave their homeland to travel across the sea to what they hope will be a better life. Throughout their journey they must overcome many hurdles to finally make a life in America.

The family settles into a little mining village where twelve-year-old Steve finds himself caught up in the everyday conversations with local towns people, who are disturbed with the deceit brought about by the different business organizations, what they are doing to the environment, and the unfair treatment of the miner and his family, as the owners were making big profits.

He tries to learn all he can about the mining industry so he can be part of solving problems faced by the miner.

As he later becomes educated as an engineer he is looked upon as an icon in the coal mining arena when he is expected to use his experience and ingenuity in saving the region from a catastrophe.

COAL
Photo from: Luzerne Memories CD, Copyrighted: Tammy Lamb

The majority of contents of this novel are fiction, however many items and events have been taken from various articles from the www.internet , and newspaper and magazine articles. Some of the experiences of the miners are actual facts of my family's life and struggles.

about the contents can be directed to S.A Wzorek @305 Stoneycreek Rd, Clarks Summit, Pa 18411 or at E-mail Norton303@aol.com

I would like to give special acknowledgement to my sister Bunny Wzorek Webb for her assistance in gathering information, writing many of the articles and doing such a great job on researching the many resources and pictures that are contained in this book.

Also my thanks to Mike Phillips for critiquing the manuscript and offering his expert opinion during this writing.

This story was written to give awareness to all of the possibilities that can occur when society attempts to change the environment and money controls the habits of the wrong individuals. Stephen A Wzorek originally formulated the theme of the story. Research and certain writings were composed by Bernadine (Bunny) Webb.

Although most of this story is fictional there is a great deal of truth in the pain and obstacles that had to be overcome in order for our families to create better lives for themselves and for us.

This book is dedicated to the memory of our father and mother Steven and Lucille Wzorek, a coal miner and his wife, our grandfather John Wzorek, also a coal miner and our grandmother Kathryn Wzorek, who worked so hard to make a better life for my sisters Bunny, Dolores, and for me.

**Steven A Wzorek Sr
Dolores, Bunny & Stephen Jr**

THE ATTRACTION

Throughout America in the late 1700's, one type of industry that flared up was the coal industry. Mining companies were formed creating jobs from coast to coast to extract the black diamond from Earth's subsurface. The coal was being mined and used for industrial use and home heating.

Different types of coal from various parts of the country were being mined: Bituminous coal (soft coal) in one area and anthracite (hard coal) primarily in the eastern part of the country. Most anthracite is used as a fuel for household heating systems. Bituminous is the most important and plentiful type of coal. It is the chief fuel in plants that generate electricity with steam.

The problems that surrounded coal mining attracted unions and other organizations. Corruption evolved throughout the corridors of the mining industry. Different tactics from all sides caused circumstances to arise, which demonstrated obvious mistrust among company officials.

Coal was being sought after from above the earth's surface, known as strip mining, and from below, known as deep mining.

Strip mining was done by construction equipment better known at the time as steam shovels that would remove hundreds of tons of dirt and rock from the surface until coal was found. Once the coal was located, it was extracted and hauled away to breakers where it was crushed, screened and sorted into various sizes. Large crater-like voids remained in its place as deep as 200 feet deep and hundreds of feet long.

Among the breakers and the holes were large mounds of leftover coal known as culm. The culm was the remains of coal dust from washing the coal at the breaker. As the coal was hoisted up the mineshaft in cars it was lifted to a higher level where it was then dumped onto shoots. The shoots had various size grates that would sift the coal. Water was flushed over the coal and the culm was what remained. It was then hauled away onto piles.

Another type of mining known as deep mining did not leave visual scarring, but the activity that occurred below the surface could pain a community for decades. Unprofessional attitudes and orchestrated criminal activity would be disastrous for coal mining towns.

Up and down the east coast, owners, or mine bosses as they were called, and union officials helped to begin episodes of tales that eventually divided the meek from the monster.

Most of the common men, those who showed up for work everyday, carried out the job with diligence and perseverance. Every man put in a full hard day and he did it without question. He did whatever it took to feed his family. He took all of the chances that were laid before him by the company, and did whatever it took to get the job done. Each man had no idea what his future could bring and he could only worry about today. He faced the hardships of a coal miner each waking day. He was aware that there was no alternative to providing for his family. The threat of losing his job was always a concern for him. He had to be careful whom he spoke to, and what he said. He was expected to be loyal to his mine boss by being on the job everyday on time, performing when the whistle blew, and to never voice a complaint.

As the men lived day to day, the company owners plotted to squeeze whatever they could out of the mining industry. As history proves, there are lacerations of every type, which linger on the face of the earth and will remain forever. There are marks of human mistreatment left with the families of those who worked the coal industry and those scars will never go away.

To this day stories are being told of the lives of local families and their neighbors who survived by giving everything of themselves to the mines and to those who ran them. The coal veins and corridors beneath the surface have taken on a new design and a new status. It is a change that would prove to be fatal to the region because of an inexcusable amount of greed and deceit that yielded within the underworld.

In the early 1900s, prosperity found its way to Northeastern Pennsylvania through the natural resource of Anthracite coal. Throughout the 1930s and 40s many entrepreneurs discovered different ways to entice the worker, who's dream could only be to live a simple life. People flocked from all over to get a piece of the action that made the valley flourish.

Politics made stride not only in the city operations, but also in the control of the mines and manufacturing operations. Every person in the valley was touched in one way or another.

The effects of the mining industry left wounds that could never be repaired. The disfigurements that remain will always be a part of the history that is conducive to the political agenda embodied in the valley of coal. It began with opportunity, followed by profit and greed. Somehow the power and politics were even stronger. This was the demise that seemed to shadow the area with misfortunes. The area always seemed to have a sort of magnetism for political opportunity. Whenever a business of any magnitude took root, the politicians had a piece of it. It also appeared that there were certain mob interactions involved in the scheme of things. There were businesses of a constructive nature and those of destruction. Anyone who was anybody was in on the destruction.

CHAPTER I

It's Just the Beginning

My parents were Polish immigrants who came to America before the beginning of the Great Depression. My father was desperate to make a good life for his family. He was a coal miner by trade.

On October 20th, 1926, with what little money they were able to save between the two of them, my parents John Wzorek, 24, and Kathryn, 23, along with their only child, 6-year-old Claire, packed their trunks with whatever belongings they could carry. Dad told me, "We only packed our bare essentials. Since we would have to pay for each bundle we took on board the ship. Each bundles size and shape was important because everything had to be carried by hand." As you can imagine, packing choices were very difficult. They left Krakow, a small city 115 miles outside of Warsaw, Poland and headed for "The America", as they called it, where they heard of the "better life." Mom once relayed to me, "We were very scared because we didn't have any other traveling companions with us." Both of them left behind relatives including brothers and sisters. They had a dream to fulfill, but they would share the title of immigrant for the rest of their lives. My mother was pregnant with me and within a short time after our arrival I was to be born.

Ships were the only means of travel for those who wished to migrate to America, and the voyage would be long and difficult.

The Estonia

Passengers: Jan and Katarzyna Zorek
Port of departure: Danzig/Gdnask
Date of arrival: November 10th, 1926

The price of a first or even a second-class ticket was unaffordable for my mom and dad so it forced them to purchase a third class ticket. In the 1920s, the competition for passengers among the steamship lines was fierce so the price of a third class ticket was cut in half to just ten dollars for an adult and half price for a child. Not many wealthy citizens emigrated from Poland. To my parents this was an enormous amount of money, but from stories they heard about the land of opportunity, this was an investment that would reap big benefits.

Throughout my life, my father and mother described to me some of the incidents and the turmoil they faced during their initial journey.

"We left everything we knew behind us." They said as they reached the port of debarkation, "We realized we would be unfamiliar to others and that we were on our own. There were few that could understand or explain what was expected of us, but we soon found out."

Before boarding their ship they had to take an antiseptic bath and have their baggage fumigated.

They were just three of at least one thousand third-class passengers aboard the ship. Unlike the first and second-class passengers whose living conditions while on board were more comfortable and luxurious, third-class passengers were directed to the bottom deck, which was also referred to as the steerage compartment (ships of earlier years had their steering mechanisms in the lower part of the ship) or "down in the hold."

As my father explained in his broken English, "We were put on board, jammed shoulder to shoulder. I couldn't turn around. There were so many of us, and the smell was terrible." The air in the steerage became rank with the heavy odor of spoiled food, seasickness, and unwashed bodies. The smell of disinfectant was overpowering and there were no windows in the bottom deck to eliminate such odors. However, the smell of disinfectant would have been refreshing compared to the odors that would build as the trip continued on day after day.

The sleeping areas were long hallways with blankets hung, dividing them down the middle. These hallways were jammed with metal-framed bunks along the walls where men slept on one side and women and children slept on the other. The mattresses were old and torn. Mom recalled, "It was difficult sleeping because of the engine noises and the ship rolling back and forth. I would just lay there restlessly, with many disturbing thoughts. Why did we leave home? What would we do in America?" My dad recalled, "The air was so thick with smoke, and your head itched with lice."

Each person was given a tin mess kit, which contained a metal dish, a spoon and a fork. They ate their meals from their beds, unlike first and second-class passengers who had tables, china, and linen. Some passengers suffered from malnutrition. The meals were of low quality and usually consisted of potatoes, eggs, stringy meats, fish, and lots of stew and soup. As the voyage progressed, the food in the steerage was cut back. "We sometimes only received a cup of soup and some dried bread," my dad said.

There were no showers or any type of bathing facilities. The only means they had for cleaning themselves was a couple of wash bowls with cold water in them. The smell of unwashed passengers got pretty bad after a while. Rough seas during a storm were unbearable. Passengers became ill and were vomiting or complaining of nausea. Without windows it became quite smothering. Once each day, the steerage passengers were allowed to leave their quarters for a walk on the deck where most of them would vomit over the railing. But if the sea was violent, they would lock and even tie the doors shut, so no one would go to the deck and be washed overboard.

Once a ship left port from other countries there was no telling where each person's final destination would be. As they traveled across the sea toward American land, they could only make a judgment as to where they would journey by the mere gossip they would hear on the ship. Different groups of people thought they knew something about a particular area in America. That was generally how they decided where they would settle

After three weeks at sea, the ship finally headed into the New York harbor with its horn blasting. Everyone cheered in his or her own ethnic language, but it all meant the same thing. They felt so much pride and joy to have reached America. They were thrilled to have the opportunity to reach yet closer to their dream.

As they disembarked the ship there was a lot of confusion. It was difficult for most of them to understand what was being said. Immigration officials and doctors who boarded the ship to inspect passengers for diseases greeted the first and second-class passengers. If they passed the visual inspection they were allowed to leave the ship.

Third-class passengers were taken off the ship and were immediately put onto a ferry, which would then transport them to Ellis Island. This would be their first glance at the Statue of Liberty. The huge statue mesmerized my dad. He had heard about it and knew that it meant hope and freedom. My dad said my mother was so overcome with emotion that she just stood there and wept.

My dad smiled as he told the story. "It took three weeks to reach America and by this time we were a raunchy looking bunch of greenhorns, reeking of body odor and smelly clothes."

Finally, the ferry docked at Ellis Island and everyone rushed off carrying their cherished possessions. Officials told them to get in line to enter the processing area and then met them at the end of the runway.

Passengers disembarking ship to enter the processing center of Ellis Island.

As they entered the processing area, a man was shouting at the top of his lungs, "Put your luggage here! Drop your luggage here! Men this way, women and children this way!" Dad told mom, "I'll meet you back here at this pile of trunks, and I hope I can find it again, I'll see you later." It was somewhat scary, mom related "I felt like I was in a herd of cattle waiting to be branded." If a passenger looked dirty, he was taken to an area to take a shower. All of their clothes were taken from them and when they got them back, the clothes were all washed, smelling fresh and neatly folded. They were assigned numbers and they were given tags to pin on their clothing. The tags had their name, the ship they traveled on, their country of origin, and a number.

They went on to the registry room. It was a huge room with high ceilings where everyone was to be registered. The room had pathways divided by railings. Along the railings were inspectors, wearing high leather boots and stiff looking colored shirts. They seemed to be watching for an assortment of physical problems.

An aerial view of Ellis Island

Stephen A. Wzorek

External view of Ellis Island with immigrants waiting to be processed

A stereogram of people leaving the ship.

A high view of the registry room showing the maze of aisles people were required to pass through.

Nighttime frontal view of Ellis Island

Passengers being examined

Stephen A. Wzorek

Ocean view picture leading to Ellis Island

Immigrants on a crowded ship

Officers and doctors were there to check for diseases. If an immigrant had a medical condition they would get a letter on their tag written with chalk. The letter symbolized a particular problem. The person with the chalk letter was then directed to a specialized doctor, who would thoroughly check them to make sure they actually did have the disease. Some were deported back to their country or detained in a medical area. The ones who made it through were given eye exams. The doctor put a hook up into the eyelid and lifted it. As anyone would expect, my parents said this procedure was terribly painful. They later learned that the exam was for Trachoma, a very serious eye disease that could lead to blindness. If you had Trachoma you were immediately deported.

The immigrants were tested for their vision, hearing, strength, and their coordination. The tests included a beanbag toss, balancing a book on their head while walking, skipping rope, and lifting barbells.

At the end of the line an inspector sat at a large desk. With the help of a translator he asked numerous questions such as, what is your name and what country are you from? He also asked my dad if he had a job promised to him by any company. My dad answered "no" to this question but had no idea at the time how critical this answer would be. He later found out that if he had said "yes", he could have been deported. There was an immigration law that did not allow contract companies to bring in jobs to America. The last question my father was asked was, do you have any money? You had to have some money so you could take a train or whatever other mode of transportation you needed to get to your destiny. My parents had twenty-five dollars, which was considered to be sufficient. So, with the twenty-five dollars and their few possessions, my parents started out for their new life in America.

After exchanging their zloty (Polish money) for American currency, they were lead to a ferry, which took them to the mainland for the first time. They couldn't take their eyes off of the Statue of Liberty as they left on the ferry. They then went to the train station to await a train that would take them to Fall River, an industrial city in Southeast Massachusetts.

Pride: Sailing away from the Statue of Liberty

Now in a strange land, with very few possessions, they began to cling to anything familiar. A few Polish speaking men and women stuck together in a small group. My parents were among them. Some had heard of many other immigrants before them, coming to Fall River, where jobs in cotton mills and shipyards were plentiful.

On December 12, 1926, after traveling for three days on train, my parents entered Fall River. My mother was seven months pregnant with me, so this had been a long journey of many emotions for her. Fear, anticipation, and now gratitude had entered into their life. I was soon to be born in America.

The group found shelter in a tenement building. The four families had separate living quarters, but shared one kitchen and an outhouse. My mom often reflects back to the memories of the outhouse. Shivering, she recalls, "Can you imagine what it was like to have to go out to the outhouse late at night or in the middle of the night, when the temperature has dropped below zero, or on cold stormy winter nights?"

"Thank goodness they are gone. Those times linger with me and I so appreciate all the modern facilities we have today" mom said.

At least they were able to communicate with each other. They also shared stories of their homeland and shared the responsibility of keeping food on the table, keeping the place clean, and making sure everyone was clothed properly.

My mom worked just as hard as my dad. While the men went job hunting, the women cooked their favorite ethnic foods, and did their other chores.

Mom says she does not miss those difficult days. One morning while starting her laundry she related to me how much harder it used to be. "Just to do the laundry we had to carry our dirty clothes to the nearby creek. We would place them on rocks so the water could beat against them. We then poured buckets of water into a galvanized washtub and boiled it. First we had to scrub the clothing on a washboard with Lye soap. We would make Lye soap from lye, which was made from ashes, water, and fat. We would take the lye water and boil it with hot fat skin and all, and make a heavy-duty cleaner. The lye soap was painful to use because it irritated my hands. We then put the clothes into the boiling water. We hung them on a line to dry after they were rinsed in another tub of water. It was hard work, but the clothes smelled so good from hanging outdoors."

The rest of the week was filled with chores such as baking fresh bread, canning fruits and vegetables, making butter, and drying meats.

Three of the men got jobs in the cotton mill and my dad was able to get jobs doing carpentry work for the town's people. He was able to keep himself busy and making money through word of mouth. He was extremely good with his hands and he was capable of building anything from a shipping crate to a house. He did a lot of home and yard repairs and even was recruited to help build a new school.

He also became obsessed with learning English. He encountered people everyday that spoke English. He wanted so much to be able to communicate back to them: he began classes to study the language.

Within two months of their arrival, on February 2, 1927, I was born and named Steven Adam

Wzorek. My dad still spoke in Polish at times with my mom and according to her, he picked me up and said, "Clopak jak swieca", an old Polish proverb meaning, "He will be a youth like a candle: tall, straight and bright."

In the morning just as I was about to be born, my mom was picking coal at the culm dump. She told me: "I put on my babushka, a covering all Polish women used to wear on their heads, and set out to bring some coal home. I had my wheelbarrow that I used for our vegetable garden to haul the coal from the dump to the house. I took a bucket and climbed the dump to get to the coal. It was a pretty steep hill to climb. I really don't know what I was thinking. When I had the wheelbarrow full with about four buckets, I headed home to crack the coal. But when I got half way home, I knew I needed to hurry because it was time for you to be born. I had to leave my coal by the side of the dirt road."

Although my father had employment, he was not making the money he had hoped he would. Compared to a miner, a carpenter's wages were meager and right now there weren't any job openings in the mines. Beside the issue of money, my dad's heart was in mining.

It was becoming extremely difficult to survive in that environment. To add to their concerns my

18

mother had to care for her second child. They barely were able to feed us. Some of the other families tried to help us, but as my father told us the stories, he said, "We became a burden to others."

To make things worse the Great Depression hit America and the times were extremely difficult for everyone. My father would have to make yet another decision to do the right thing for his family. He admitted sadly, "I am so fed up with working so hard with such low pay, day in and day out, at something I really do not want to do for the rest of my life, and not making enough money to keep us going. Things are getting worse and I am not able to make a future for our children. I have to give up and go back to see what it's like at home now."

Mother, in disbelief questioned dad's decision: "Who knows what it is like back there now. It could be worse than here. What will we do after the money we saved runs out? Where will we live? Where will you get a decent job?" Dad answered: "All of our family is in Poland. Families stick together, and to tell you the truth, I miss my brothers and sisters, don't you?" "Of course I do", mom said, "but think about the children. We came here to make a better life for them!" Dads logic was, "We will take the money we have saved from America and live a richer life in Poland. We can also help our families get out of debt. Then we will come back to

America with the children when I am sure of getting a good job maybe in the mines or something like that."

He would have to take whatever they had left and head back to Warsaw, Poland and Krakow. He knew he could not make it under these circumstances.

After a few months of fruitless effort we faced the journey back to Poland to be with the rest of the family once again.

My father had no intention of giving up. He always said he would go back to America once more to live out his dream. He remained in Krakow to once again work the mines.

He was continuously telling his friends he was going back to America. On the way home from the mines after a hard days work he would speak of America. They would laugh and make jokes about him attempting the trip again. They asked him how he would do it with yet another child, and it would be more expensive. He said, "You will see, I will do it. I live every day to make that journey back. You will see. Someday I will go back to America."

Everyday as they joked with him it just inspired him more. He worked harder each day and asked the mine boss for more work time to save more money.

My mother said each day he would come home and try to figure how much he needed to pay for the trip. She said he talked to all of his working buddies because he was looking for anyone that could help him. He wanted to know if they had friends or relatives there. It seemed all of his family and friends in Krakow knew he was going sooner or later. He hoped to find someone in the Northeastern area of America.

They saved enough money to attempt the trip back to America once more. However, this time he did some research and learned of an industry that was hitting the northeastern United States and the East Coast. The coal mining industry was taking over. It was making headlines throughout the world. It was an industry that he was familiar with.

Stephen A. Wzorek

In this May 1956 photograph, people stand on an excavating machine called a dragline at a strip-mining operation on Coxe land in Pennsylvania.

Strip mines use huge excavators to remove earth and rock covering coal seams.

CHAPTER II

After years of searching, my Dad finally met a family by the name of Ochwat. They had a brother that four years earlier was able to find his way into the hills of Pennsylvania where the coal was. This was ideal for my father. He was able to get the address of Stanley "Cookie" Ochwat. He wrote, asking him to help him find any work available for a coalminer with his experience. Cookie wrote back and said he could get him work whenever he got there.

Cookie wrote to my father: "My dear friend John, I am so happy to hear from you. My sister Irene told me all about you and your family and I am looking forward to meeting you upon your arrival to America. I understand you are not a stranger to our land. I have been here for four years and I am doing well. I was able to find work in a small town here in Pennsylvania. This area has a lot of work and there are a lot of shafts opening all the time. There are other towns nearby as well. I know the mining boss and I will talk to him about you. There are many nice people here. Some from Ireland, Whales, Italy, and some, mostly from Poland. Some of us are still having a hard time talking to each other. I think you will fit in very well. I will let my friends know you are coming and we will help you get work. Let me know when you are coming and I will make arrangements."

From that time on, letters were sent back and forth between the two of them.

Cookie was living in the small mining town just south of Scranton, Pennsylvania. It was about one hundred thirty miles from New York.

The times had changed. By law, anyone wishing to leave Poland was required to obtain exit papers. To ease the process of admission to America, a letter from family or friends in America promising care and employment for the new immigrant was very desirable. My dad received his request for immigration and obtained a letter of welcome from Cookie in America.

They finally set a date when they would meet. Cookie said he could have transportation for my father, mother, and the two children when they got to Scranton.

We began to make our return trip to America on June 1st, 1937. I was eleven years old at the time. My Dad kept checking the prices of the tickets. The cheapest ticket was seventeen dollars, as opposed to ten dollars, ten years ago.

As we approached the dock of the Trojan ship, I wondered if it would be the same as it was for my parents on their first voyage by sea. I hoped it would not be as bad as the stories they had told me.

The Trojan ship was an American-made vessel. When we entered the ship, Dad, looking somewhat relieved, told me, "This ship is much bigger, faster, and more beautiful." Her atmosphere was simple but still warm and pleasing. I noticed a lot of aluminum and shinny stainless steel throughout the decor. The American ship lines shifted from carrying their immigrants using crude steerage hold, to accommodating them in "new steerage". Rather than the older, more crowded, noisy, odorous, and immodest open bunkrooms, characteristic of the old steerage; the new ship had private closed individual berths for each passenger.

We had three meals a day, cooked on a wood stove and served buffet style on the deck. The men had separate compartments from women, with towels and soap for washing onboard.

Not very many people complained of seasickness. Mom said, "Maybe because the ship is roomier, has more deck space, and more fresh air."

The seas remained calm through the entire journey and the trip back to America this time was much more enjoyable for everyone.

Immigration patterns changed over the years. Dad was so well prepared he had all the necessary paperwork and forms we needed to shuffle us through immigration. He proudly presented my birth certificate and passports for everyone else. We were processed onboard this time and never stepped one foot on Ellis Island. At the time, Ellis Island was used as an administrative and detection facility due to the effects of World War II. Only those immigrants held for hearings went through Ellis Island.

This was a big day for all of us. My mother, father, sister and I arrived in New York on June 18th, 1937.

We left the shores of Ellis Island by train and headed for Pennsylvania. We traveled on a train that was known as the Phoebe Snow. It had fine dinning cars and sleeping berths and we were treated as well as anyone would expect. Two days later we entered the station in Scranton, Pennsylvania, where Cookie and his family met us. Cookie, his wife Mary, and their daughter and son were standing on the platform of the train station. They held a big sign with our name, Wzorek, on it. It was so inviting. My Dad and Cookie greeted each other in Polish, "Yak she mash", (Hello, glad to meet you). He said, "It's a pleasure to finally see you face to face John." They spoke to each other for a few minutes, and then we piled into his car and were on our way. It was really tight with all of us in the car. The luggage was all tied to the roof. We were headed to Minerhill, a small village surrounded by coalmines.

As we approached the little town, there were clusters of houses lined up on each side of the street.

Stephen A. Wzorek

Every house had a small vegetable garden planted in the back of it.

Company homes

Cookie informed us, "Minerhill is known as a 'patch town', because of the patches of gardens in every yard."

Cookie lived in one of the houses. He told us that he was able to make a deal with the owner, his boss, to agree to provide my parents with a home in exchange for a job in his company. This was a regular occurrence because so many coal miners were urgently needed in this colliery. In fact, miners were discouraged or sometimes prohibited from living elsewhere. The boarding house was there for my dad; all he had to do was agree to take the job. The houses were small and offered no privacy. They were made of hickory wood, which was available and cheap and as a result it attracted bed bugs. Some of the houses had dirt floors and the walls were stuffed with rags to keep the heat in. My dad was able to lay a nice wooden floor. He also patched all of the cracks and holes in the walls. Although my father tried to make the house more comfortable, it always smelled with dampness and stale coal soot. We took a bath in the kitchen once a week. My mom would set a tub along side of the stove, heat water and fill the big tin tub. We all used the same bath water. I bathe first, then my sister, then mom, and dad was last. There were only three rooms. We slept so closely together in rooms next to each other that it was hard to get good nights sleep. My dad would cough most of the night from swallowing so much coal dust.

Most of these miners' shacks in the 1920's did not
have electricity , running water, or sewers.

**This one room shack has roofing material used
as siding and needs a screen door fixed.**

Large families sometimes lived in these one-room shanties and there was a complete lack of modern conveniences. A water barrel collects rainwater.

I vaguely remember the house but it had two bedrooms sparsely furnished and the kitchen had a big coal stove. I remember my mom putting four bricks on the stove to warm every day. She then put them in our beds before we went to sleep. The house had a front porch that I loved playing on. After our first day in our new environment we settled into the house. The next day Cookie came to take my dad to meet the coal boss.

After several hours with Cookie, and finding out where he would be reporting, Dad came home and walked into the kitchen. He nodded his head saying, "Bez pracy niema", meaning, "Without work there is no bread." He had once again taken the job of a miner. His pay was five cents an hour, double what he made in Poland, and he would work twelve hours a day.

I helped my mother plant a vegetable garden to help feed the family and to trade with the town's people for other staples. The days started early. School was out on summer break, but we would get up 6:00am. The birds would be chirping, and the fresh air smelled so good. We brought some of our last year seeds we had from Poland and began planting. We planted corn, potatoes, carrots, tomatoes, green beans, and onions. We also planted wheat for flour. After eating supper, we were so tired we went right to bed.

Cookie and my dad became very good friends. Cookie was also a miner in the colliery, and he was a straight to the point kind of guy, who pulled no punches. He told it like it was, and didn't care who or what he offended. He always spoke the truth.

There were a lot of kids in the neighborhood to play with. Some were playing marbles, some were playing hopscotch and some of them were playing kick the can. Everyone was invited to participate.

I soon became good friends with one boy in the neighborhood in particular by the name of Marty Kloss. We realized we were the same age and had a lot in common. Like me, he also was a descendent of a Polish immigrant and he was born here in America. Marty became my best friend. We played together from then on and he always watched out for me. He even taught me a lot of the English language and we spent a lot of time playing and laughing together.

Cookie took my parents, my sister, and me on our first trip to the company store to pick up staples and supplies. He complained in his characteristic wit: "The company store gives it to you on the book. Every pay day they'll take a bunch off you, they take it off all the miners."

One day Marty and I were walking through the town. The coal dust lingered throughout the neighborhood. We saw women beating rugs, washing windows, and scrubbing porches to rid them of the grime and dust of the coal. Although they knew it was a never-ending battle, they tried to keep things as clean and neat as possible. While we were walking we could see some miners ending their workday. They would walk to the beer gardens for their daily after-work beer, and, like Cookie said, to "shoot the shit." Their faces were black with only the whites of their eyes showing.

I noticed there were a lot of beer gardens throughout the town. If the doors were open, you could hear accordion music and men singing, or see them playing shuffleboard. I thought out loud,

"Wow, good place to come on Halloween." We came to the grove where every summer weekend there was a picnic. I can still imagine the aroma of chicken barbecuing on the grill and the sweet corn steaming.

Being with the coal crackers, my dad began learning more and more English and he would teach it to my mother. I started at a new school, Woodrow Wilson School, on the top of the hill. It wasn't long before I could earn my way to becoming the American that I was. After all I was born in America eleven years ago.

The day before I started my first day of school, my dad motivated me to go, by telling me of how he learned the language. He said, "Stevie, I learned English the way everybody else does - by going to classes. Everyday after work I attended a special program with English classes. In these classes, I learned to pronounce English sounds. I made a lot of mistakes and my pronunciation was bad, but I had a good teacher and she worked with me. My American classmates also helped me on weekends. I got a good dictionary. I learned every new English word that I came across. With each new word I felt closer to my goal. I would go to movies. I had fun and I learned at the same time. I would go several times to a movie if I liked it, just to learn the dialogue. From movies, I got a lot of sentences to memorize. I would sometimes imitate the actor, saying the sentences like they did. Of course my English was not too good and I spoke slowly, but the movies helped me to improve my pronunciation, grammar, and vocabulary. I read many books and I got better. I am amazed and thrilled with my success. Now, another dream had come true."

I played, worked, and lived in coal throughout my childhood. I visited the mines and shafts. Sometimes my dad drove a truck for the colliery. Every chance I got I rode in the truck with him. I learned the mechanics of the trucks, the uses of different mining equipment, and the titles given to bosses and other miners pertaining to the specific job they performed.

This excavator has a wheel, left, that cuts away 3.500 tons of waste an hour, and piles it over 420 feet away.

When the mine was not operating, Marty and I, although forbidden, would swim in the wash chute at the bottom of the shaft. The chute was made of wood, shaped like a trough, and about four- feet wide and forty- feet long. It was used to wash coal by natural runoff water from the walls of the shaft. The water always flowed through the shaft, and we would dam up the trough at the bottom so we could swim in it. If we weren't swimming we were exploring the shafts and the surrounding structures. To this day I have a collection of steam whistles, transformer detectors, and padlocks, all of which I found while we were playing near the supply houses. I kept my treasures in special boxes in our house. In the years to come these treasures would hold the answer to the perils of greed.

Open lamps

Stephen A. Wzorek

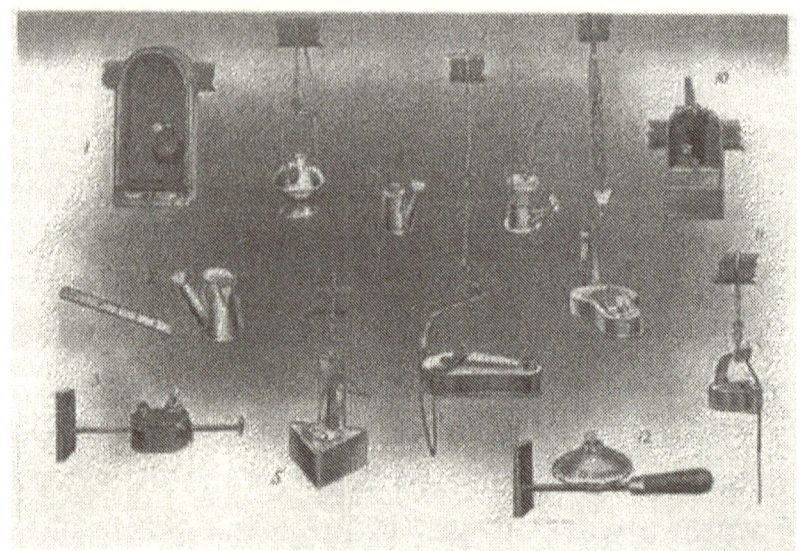

Various open type lamps used before the invention of the Davy lamp.

Ewin Collier, Pennsylvania Coal Company. (Courtesy of Joe Keating, Plymouth, Pennsylvania.)

Stephen A. Wzorek

View of a working colliery

CHAPTER III

My Home Town

A few years later, my sister Laura and my brother Matthew were born. My father was somehow able to still save for our future. He was able to purchase some land in the lower part of Minerhill on the main road, later known as Davis Street. Over the next few years Matthew and I helped my father build a three-bedroom house. We also included another room on the first floor that we set up as a grocery store for the neighborhood. This store would be different from the company store that many others patronized.

Unlike the company store, the miners didn't need script to buy supplies. Script was an IOU issued by the mining company taken from wages, and the price of the supplies was always inflated. Our store was friendly and carried credit for those in Minerhill. The people would pay when and however they could, but everyone always paid their bill. After a while we also installed a gasoline pump, and we were eventually able to supply most of the people what they needed. As I recall, we sold a lot of Fels Napta Soap. It was a laundry soap that helped remove the coal dust and grime from your body. It seemed all miners used it.

This illustrates typical home conditions at the turn of the 20th century. The miner returned home dirty and used a tin bath in the living room. The water would be boiled using the coal fire by his wife. A curtain is draped over a clothes maid to separate him from the children. He uses a bristle brush and Fels Napta soap to rid the grime.

Washing away the dirt and coal dust in the tin bath at home in front of the fire.

My father would constantly carry on conversations with the people who came into the store. They always talked about the mines and the problems that occurred. Throughout the day and part of the evening my mother operated the store. She, too, would get into conversation about the company and what they were doing to the workers. I remember the evening Cookie came stomping into the store. I never saw him looking so angry. "That's it, I can't take it any longer! Nobody wants to listen to me! Nobody cares what happens to me, or any of us! Dad said, "Slow down, Cookie,

What happened?" Cookie, breathing heavily, grumbled, "I am really scared John, working in that tunnel. There's a section of the roof that is going to collapse any day. The pillars are so stressed they're caving. To begin with, the tunnel is so small, sometimes I have to get on my belly and crawl into it, and dig lying on my stomach. Anytime under that roof is too long! I have been complaining to the foreman for over two weeks, and today he told me, he and the engineers have inspected the pillars and there is nothing to worry about. I am not stupid John, I can't handle this crap!"

There wasn't anything new that they would talk about. It seemed there was a general practice of abuse throughout the industry. It affected the miners all over. Only now it was happening here. Most of the people that lived in Minerhill were immigrants. Many were from Poland, Ireland, Wales and Italy. The mines have been good to our families but over the years there was a new generation of business people who wanted more as the times were visibly changing.

Billions of dollars were being made by the coal tycoons through their evils, which arose from the erroneous methods of mining.

Truck hauling companies were hired to take the coal to its destination. The tycoons eventually bought and owned these hauling companies under different operating names. There was a lot of competition between the mine owners. Although they had claimed different sections of the land veins, they tried crossing into other owners territories, therefore causing feuds when it was discovered. The mine owners instructed the foremen to have the workers steal the coal pillars and ceilings, leaving dangerous subsurface that could later cause cave-ins.

This explosion sketch conveys the horror of an underground explosion.

A Miners Prayer

O Lord after I have worked my last day and come out of the earth and have placed my feet on Thy footstool let me use the tools of prudence, faith hope and charity. From now on until I will be called to sign my last payroll, make all the cables in the machinery strong with Thy love. Supply all the gangways, lopes and chambers with the pure air of Thy grace and let the light of hope be my guidance, and when my last picking and shoveling is done, may my last car be full of Thy grace and give me the Holy Bible for my last shift, so that Thou, the General Superintendent of all collieries can say: "Well done, thou good, faithful miner come and sign the payroll and receive the check of eternal happiness." Amen

Eyeing only the bottom line, mine owners disregarded the miner's safety. This was a common gripe among the miners. Coalmines proved the scene of multiple calamities, a direct result of the selfishness of the coal barons. Always two variables were involved in mine accidents. One was the system, or lack of a system, to have proper water drainage. The other was a proper ventilation system to vent the methane gases, sometimes present in mines.

Methane is a colorless and odorless gas. It is not toxic; the immediate health hazard is that it may cause thermal burns, but it is flammable and may form mixtures with air that are flammable or explosive. The methane would accumulate and explode as it often did. If the drainage system or pumps malfunctioned, there was severe flooding. More dangerous was the ventilating furnace that would catch on fire and burn the support timbers, subsequently depleting the oxygen in the shaft and suffocating the miners.

This was exactly what the miners feared. If the ventilating furnace set fire to the woodwork in the shaft, hundreds of men and boys could be killed. When an accident like this occurred the coal companies and their officials remained virtually blameless; instead they blamed workers for their careless disregard for safety. A proper ventilation system in the mine such as fresh airshafts could just about eliminate the hazardous effects of the methane gas. But for this matter, the coal baron had no money to spare.

The dangers down in the hole were many. The miner could be crushed to death by a falling roof, burned to death by an explosion, or blown to pieces by a premature blast. It is to the benefit of the coal baron to work the seams completely out as much as possible. This practice causes the roof to cave in, burying or critically injuring the worker. The coal bosses incompetent and ruthless pressure for increased coal production left a sad litany of problems such as inadequate ventilation, poorly constructed shafts, and no safety management training. There was a rumor circulating among the miners that Minerhill Colliery and some union officials had established a sweetheart contract that pushed safety aside. The deep-pocketed politicians blindly supported this risky venture. This was just a small part of the complaints that the miners mussed over.

I remember the day I was working the gasoline pump at the family store. The colliery whistle blew at an abnormal time. In about twenty minutes all of the children of Mrs. Rozaieski, one of our neighbors and a wife of a miner, had come to her home on our street. I watched as they went back up the street in various vehicles. If the colliery whistle sounded and it was not for a shift change, you knew there was a cave-in or an explosion.

The next day was warm and sunny. The colliery had shut down. (If there was a fatal accident the shaft closed for three days). It was a day of sadness. People began to gather in the store, telling of their findings. A gas explosion killed Mr. Rozaieski. He went into a chute with a naked (uncovered) light, igniting the gas and fatally burning himself.

Mrs. Rozaieski's brother, Stanley Lisiewski, a miner also, blamed the mining industry and the coal bosses, "There are no professional mining engineers working this colliery. The bosses depend on miners to operate the colliery through practice. We need to make these shafts safer. Forget about those greedy sons' bitches and their big bucks." Dad interrupted: "If there is an uncovered light you're sure to have an accident." Jim Heenan, an Irishman and good friend to Mr. Rozaieski stated: "Well then, don't have a light in the way. Make the walk in the dark!"

"That's easy for you to say," remarked Stanley, "Now that's going to be a job, getting the men to do it in the dark." "But if it was all done in the dark," my dad questioned, "what if a boy came running through the shaft with a lighted candle whistling Dixie, then what? Then where are you?"

Jim, sitting on an orange crate, puffing on his pipe, looked at Stanley and said, "You say there is no authority or order down in the mines, nobody walking the shaft to keep these kinds of accidents from happening because of miner's carelessness. Well, there's the foreman giving out the work. Then there's the super who is always walking about seeing that the workers did their job. What about them?"

Stanley, shaking his head wondered aloud, "But what's one man to miles and miles of darkness underground with gas and bad air everywhere, and roofs and walls that could fall in at any time?" To this dad replied, "I see no help for accidents from explosions now, or ever. As far as man is concerned, I myself have been careless many times, as I'm sure all miners are, and always will be. You might be able to cure the mine of the gas maybe, but you'll never cure the miners of accidents. I can't really see how you're going to get rid of all the gas either."

My mom felt bad about the miner's wives and how the families were treated. She began to tell of so many miners' wives telling their story of how the mining officials mistreated them. All the wives of injured or killed miners told the same story of how the mining company would bring their loved one and put him on the kitchen floor. The poor wife felt helpless, not knowing what to do. She usually had no money. The courts held that the state was responsible for the mining accidents and not the company so families rarely filed damage suits. Neither the state nor the company offered the survivors any compensation. Some companies offered the survivors $100 and a return ticket back to their country of origin.

In trying to get the conditions safer, the miners decided that they would strike. For every little thing there was a strike. The guy with a grievance would pick up his dinner can; spill the water out saying, "Domu" (home), and all of the miners would get up and go home. They stayed out on strike until their conditions were met.

Firemen carry canaries down the pit. Under normal conditions a canary sings continually. When gas is too dense the canary stops singing. This is a signal to miners to withdraw from the danger.

A miner drilling into the seam. This photo shows how hard the work is for a face worker, the heavy drill, the noise, dust, lack of headroom, kneeling on the rough ground, and the constant danger.

In the days before the arrival of the coal-cutting machine, a collier can be seen undercutting the coal with a pick. A lamp hanging on a prop was his only indicator of the presence of dangerous gas and his only source of illumination, about a half candlepower.

Preparing to blast, the miners below, use a pole to tamp the explosive charges into place in the holes.

A shot-firer, as the miner above is called, prepares an explosive charge by inserting a detonating device.

Props: From the early days of mining, wooden posts or pit props were used to support the roof. These had to be removed as the coal face advanced. This photo illustrates one man cutting the coal with a pick and another man setting a prop to support the roof.

Many mine barons and company bosses pushed the men to their limits, constantly trying to get more out of them. The more the miner gave to the mine and the bosses, the less he received in return. Half of the streets and all of the shafts were named after the coal barons. They never left anything for the miner and his family, or for that matter the community as a whole.

As we lived the life of coal mining families we continuously heard of different organizations. It wasn't long before more mining proponents were filtering into Minerhill.

The effects of the Molly Maguire's from the past years, the Unions and tough competition, continuously affected the business ethics of the daily operation. Trust was still a major issue between the owners and the laborers.

The Molly Maguire's were a secret organization dating back before 1870, which was made up of predominantly Irish Catholic men. Their intentions were to focus on regaining control of the mine owners and superintendents in their struggle against injustice. The Molly Maguire's and all mine workers rebelled against the wealthy mine owners because of the conditions they were subjected to in the mines, and in their daily lives. In 1870, twenty-seven alleged members were tried, convicted, and hanged for the murders of a number of overbearing mine foremen and owners.

Whether the convicted men actually committed the murders remains a subject of some debate in the anthracite region today, especially because the prosecution was paid for and organized by the area's dominant coal companies. There is no doubt that the Irish mine workers were responding to what they believed was autocratic and oppressive control over the mineworkers and managers. The Molly Maguire hangings ended the first wave of violence in the coal region. Labor relations throughout the United States remained turbulent, however, and the battles between mine owners and mine workers continued. Frustration on both sides led to violence through intimidation, beatings, industrial sabotage, and military intervention.

Violent criminal acts were constantly occurring in the mining communities from as far south as Schuylkill County, Pottsville and Tamaqua, and north to Carbondale. Highly publicized murders have regularly occurred throughout the years throughout the valley.

With working conditions as bad as they were, it didn't take too much for the miner to consider striking. As the coal industry bloomed, the strikes occurred more frequently, resulting in vicious attacks, mostly on the miner. The coal bosses used physical or verbal threats to intimidate the striker.

Some who honored the strike had their belongings thrown on the sidewalk by the coal company, and were evicted from the company home. Others who chose to go to work were beaten and harassed by the strikers. Large groups of strikers would go to the site where the worker was and threaten him with bodily harm if he did not stop working. Many of the men on their way to work would be turned back by mobs. Some were forcibly pulled from the street, beaten, kicked, and dragged by their heels, then left for dead. The violence sometimes spread to the families. Mobs made late night visits to houses. If they knew the man was away they would threaten the wife with blowing up their home if she didn't convince her husband to go back to work.

In the early spring of 1939, my dad lost two good buddies to this violence. My dad along with other non-union coal miners in Minerhill began an effort to join the United Mine Workers. The coal operators became alarmed by the union activity and locked out the miners. As a result the immigrant miners decided to strike. The miners wanted higher wages and their grievances to be heard. Coercion to buy from the company store at inflated prices and unfavorable working conditions also contributed to the strike. A group of miners walked peacefully, toward the Minerhill Colliery to prompt other fellow miners to join in the strike.

The mine company police encountered them near the mine and ordered the group to break up and return home. Some of them did not understand English. When they did not respond to the demand, without warning the lawmen raised their rifles, aimed at the strikers, and fired, killing nine men and critically wounding four.

They kicked and beat the dying men who were lying on the ground. The coal companies had the police, press, courts, and most of the mainstream society on their side leaving the striker to struggle alone. They fought for the rights of working people, but lost the battle so many times.

The strikes also affected the local businesses. Merchants were warned by committees of strikers not to sell anything to the working man or his family.

Chapter III

The Education

As I was growing up I learned more and more about the mines.

I listened to every word my father shared with me about his goal to live a better life and his concerns for my future. Even though I was only twelve, I began searching for the education that would fit my interests. I had the need to have an association with the mining industry, so I did research on engineering. Years later in my life I found the niche I was looking for. I discovered many curriculums involving the mining industry. I chose civil engineering with a great deal of involvement in all types of mining. Civil engineering would give me the opportunity to venture into the many aspects of mining, I had no idea existed. I would be exposed to the different types of mining, not only coal. The fact remained that even though the minerals varied, the processes were basically the same.

I was also looking for a career where I could help resolve the complaints I heard from my Dad and the other miners.

I tried to listen to every person that walked through our business as if I worked those mines myself. At this time, I was just a little over twelve years old and I would have wanted to work in the mines with my father, but I knew he would not allow it. He worked so hard and he knew so much about the mines. He eventually was made a foreman for all of Lackawanna and Luzerne County. He had a lot of stories as he traveled in and out of the mines daily for years. I was interested in hearing from him, or anyone else, every word spoken about the mines. I was hungry for as much information as I could get about the mines.

My father was given copies of every map from every mine that was to be excavated. He brought copies of the maps home and would lay them on the kitchen table. He would talk to himself as he was looking at the map. I don't think that he realized it, but I was listening to him. After he was finished looking at the map I would start asking questions. According to these maps, the mines crisscrossed the valleys with shafts and tunnel from Schuylkill Valley to Carbondale, offering access to underground workings, some as deep as twelve hundred feet. The lines on the map were interweaving and connecting, sort of like a giant maze. I studied these maps as if I was working in the mine myself. The maps showed corridors and workrooms in the mines, as well as the surface coordinates, such as the township boundaries, any rivers, lakes, creeks, etc, that were above the mines. They pointed out mine properties with mined out areas, including shaft locations, as well as areas of subsidence, due to the mining of the coal.

Now I could see the whole picture before me. Already impressed I learned every emergency exit, walkway, and mine opening that I thought I could walk these mines in the dark.

My friend Marty and I would act out the duties of the miners who did the mining. We would wear our fathers' helmets with carbine lamps mounted on them and we would go down to the basement of our house to play. We read the maps as well as we could understand them. We pretended to extract the coal with our matchbox size equipment made from popsicle sticks and whatever small toy trucks we had. The entire basement floor in our house was dirt, so we had plenty of areas to dig in. We used small pieces of coal from the houses coal bin to haul around in the basement. We used the maps to layout the areas to dig. I knew those maps like they were rooms in my house.

When my father told me he didn't need the copy of the maps any more he gave them to me. I kept all the maps in my special box in my room, and I would read them whenever I could. When I couldn't understand something I asked my father. He had all types of pencil marks on most of the maps. He said it was not something he wanted to discuss, which made me even more curious.

I held an interest in mining all the way through high school. I read any thing that had to do with the mines. My father begged me to find other interests. He didn't want me to have any part of the business.

In the company store one night, as I was helping my dad close for the night, I asked him why he was so against me working in the mines.

"Stevie", as my father called me, "If you want to know all you can about the mines, there are many ways you can keep the interest but I don't want you working in them the way I have. I don't want you to be involved in all the lies and deceit." I asked him to tell me what he meant, and he said someday I would understand all of this. He said, "Son, I am forty years old and I lived in a coal region all my life. Twenty of these years were spent working in and around the mines. Your grandfather was a miner and he died of asthma. Your uncle Matt and your uncle Eddie were miners; none of us had an opportunity to get a full education. We went to school, if you want to call it school, until we were around twelve- years old and then we had no choice but to work in the mines, screening coal and picking slate at the breaker. From there we went inside the mine as driver boys, and as we got stronger we were put to work as laborers, where we helped the miner, until we eventually became the miner. Stevie, I lost my brother to the mines, an uncle to you that you never really knew. He got buried when tons of dirt caved in on top of him. He was killed two weeks after he got his job as a miner and a month before his son was born. Son, I am only forty- years old, but look at the marks on my body. My muscles are no stronger or harder than the calluses on my hands. Look at the lines on my forehead from so much worry. Look at the gray hair on my head. Look at me! I look ten years older than I really am. You want to know why? Because every single day from Monday morning to Saturday evening, from sunup 'til the sunsets, I am in underground workings of the coal mine. I go down and have to walk the darkened chambers, sometimes for miles, till I get to the shaft on the top of the slope. On both sides all I see are logs that support the sides and keep the roof from falling down. That is something that could happen any time. Just as so many have, a cave-in could also crush me to death. The water is sometimes a foot deep as I walk the shaft. The water seeps in from the wall seams into the ditches along the walkways. If it's not the water you got to worry about then it's the coal gas blowing you up into pieces like fireworks. My everyday life is not very pleasant. When

I get up in the morning and put on my coveralls, I think of all the dangers that threaten my life and others."

Stevie, "you get old so fast from the worrying, fear, the damp gases, coal dust, and all that bad air. Not to speak of all of the cheating and lying from the bosses. I have gone through it all. I married your mother when I was seventeen years old. She was like me, born to a coal miner's family. Her chances for an education weren't any better than mine. But she did learn how to keep a family and household on a certain amount of money. We had $175.00 in savings when we got married and spent that on furnishings for housekeeping. Back then there wasn't that much work available and by the time your sister was born we really didn't know how we would make it."

"Son, this was my hope and dream to come to America. You can be different. You can do it all. You can be an engineer."

CHAPTER IV

The Damage begins

In the 1960's, what once were unethical business practices took their toll on the lives of those still living in the area.

Subsidence began occurring all over Minerhill and the region. Many abandoned mines burrowed under heavily- populated areas, which increased the potential for damage to homes and other structures. Sections of buildings tilting, small spider webbed cracks in foundations, basements of homes and garage floors began to split, evidenced subsidence. My father had cables with turnbuckles on them attached to our house, as did many other homeowners. He gave them a turn when a subsidence occurred, to keep the house from shifting. He had to frequently jack up the furnace to keep it aligned with the pipes, so they wouldn't crack, as the cellar floor sunk.

Subsidence pit behind a residence. This pit was 35 feet in diameter and 25 feet deep. (Photo courtesy of Ohio Division of Mines and Reclamation.)

This gigantic mine roof collapse occurred in July 1967 and took several houses with it. Houses in the center of the cave-in settled down with the rest of the land around them and were largely undamaged, but who wants a house at the bottom of a hole.

Minerhill residents were used to these land subsidences. Many holes had been caused by loose soil giving way underneath them since coal mining began.

My family was taking a ride one summer afternoon, just to look at the scenery. We were driving along a creek, breathing in the fresh air through open windows. All of a sudden we heard a thunderous noise and a loud swish of water. Looking in terror we saw a subsidence open on the bank of the creek, less than fifty feet from where we were stopped. The water from the creek ran into the opened hole, then under the highway, collapsing it and swallowing four vehicles. I couldn't believe my eyes. I had never seen anything like this. I was afraid to go near the hole, but my dad did, shakily I followed. We helped the people out of the muddy waters as they screamed in fear. The hole opened up to about twenty feet by twenty-four feet, with a fifteen-foot drop to the mine roof. At the bottom of the hole it spread out over an area about twenty by forty feet. This subsidence was less than one hundred feet from Route 91, a major highway. My dad angrily said, "Mother Nature surely isn't responsible for this, it's plain negligence of the mine owners." On lookers began to crowd the creek road, gazing unbelievingly at the situation. It took us quite awhile to regain our senses after witnessing this catastrophe.

Driving somewhat erratically along, we saw mountain after mountain of culm dumps. Dad explained to us why there were so many. "There were millions of tons of coal extracted from the mines. These are mountains of the coal waste left. They are called colliery dumps, or culm dumps. These piles contain a mixture of coal, rock, and shale removed during processing in the breakers. Some piles contain extractable coal, while others consist of only rock and shale." Mom sorrowfully said, "What a shame, the natural beauty of the mountains are now surrounded by what looks like craters and potholes on an abandoned planet." My brother Matthew complained about the smell. "It smells like rotten eggs." "My dad told him, "That is the coal emitting polluting gases like sulfur and carbon. These dumps sometimes burn for years and at night you can see the embers of blue and yellow for miles away." He continued, "Sometimes people throw a match to garbage that they bring to these dumps and start it on fire. The fire eventually burns out of control." We stopped to watch a team of coal company contractors try to remove a mountain of the burning culm. They had already erected a large temporary pond near the dump. They used earth-moving machines that my dad called euclid to load the culm and then ran the load and the euclid through the pond so the culm would get soaked and quenched.

Dad said," The evidence of mining will stay with us generation after generation. It's never going to go away." I was now more determined than ever to get involved in solving these problems left by mine companies." I have to say, "It was a very terrifying, but interesting Sunday drive."

Pond for washing coal and discarding waste

Everyday brought a new disaster. I can't forget the day Joe Mancheski, a miner friend of my dads, came into the store in hysterics, screaming, "You have to come and help." My dad, mom, sister Laura, my brother Matthew, a few customers and I ran out the door into pouring rain, frantically trying to see what he was talking about. He was yelling so loudly, neighbors ran out of their houses. We followed Joe up the hill to where he lived. On the other side of the hill was desolation. The lucky people lived on top of the hill.

Also on top of the hill sat Coopers Coal Company. The company began dumping waste from the mine at the head of Miners Creek, right near the edge of the creek. The mine owners had cut down trees, ruined the plant life, and eroded the topsoil to get at the coal seam beneath.

The creek was already filled with sludge and rocks that washed off the hills. These culm dumps grew and grew until they began to block the creeks flow below the mountains. The company didn't install any type of drainage ditch to release water from the creek if too much pressure built up. Instead, they figured the water would find its way to eventually trickle through the culm dumps and avoid buildup to the creek. I don't remember the name of the creek, but I do remember it was a pretty swollen creek

Consequently, it rained heavily for a week. The barren hills could no longer absorb the rains. The creek, which was already at its capacity, rose rapidly. The water backed up behind the culm dumps. Even though the creek was rising and company officials feared the culm dumps might collapse, the company continued pumping water from their cleaning plant back into the swollen creek at a rate of two hundred gallons a minute. They did nothing to warn the town's people of the imposing danger other than warning a few families that the dumps might give way. They had done this so many times before that the warning went unheeded.

Only when the first dump began to break loose, did the company officials call for a bulldozer to dig a drainage ditch to relieve the pressure of the water on the dumps. The company officials thought the drainage ditch would prevent the dump from collapsing, thus failing to inform the Department of Mining or the local police of the eminent danger. Ultimately, the culm dumps completely collapsed one after another, spewing the black coal waste into the creek water. Massive explosions were set off as the rushing waters were met with smoldering yellow sulfur coal waste deposits. A crater ten- feet deep in places was a result of more than five thousand gallons of water and hundreds of coal waste thrashing down the valley in a wall sometimes eight feet high, ripping homes and lives out of existence. Thirty homes were demolished. Countless people were injured or killed by the swirling waters. My dad, Joe, I, and some other miners helped pulled bodies out of the swirling waters. This was one more heartbreaking event connected to the mining barons.

The mines were certainly a danger, but being a teenager I guess I thought I was untouchable and was very risky. There was a bootlegging (illegal) mine behind our house, where Marty, I and some other friends would crawl through the mine shafts, all the time hoping the walls and roof wouldn't collapse on us. It smelled of damp, stale air, and your clammy skin itched from the dirt and dust. Marty commented, "How could anyone stay in these caves all day long?" I answered, "I guess this is what my dad meant when he said he didn't want me working in the mines." There were rats running all over the place. It was a horrible place to be for a little while, never mind all day. Our parents always knew when we were crawling through the mines, because we would come home smelling like old sweaty socks.

Matthew would come to the mines with us occasionally, when we tolerated him. One time he came out yelling, "There's a ghost in there." When I went into investigate this ghost, I saw two glowing eyes. The ghost, running out of the hole, turned out to be a scared fox. It was pretty funny.

It wasn't always bad living around the mines, but on a whole the mines took more than they gave.

Chapter V

A NEW LIFE

Years went by, I graduated high school. The war was breaking out, and I decided to join the military service, after all, I was an American. I became a United States Marine. After basic training I began to study engineering as part of my duty.

As the war was going on I was learning more about all types of engineering. I never took my focus off mining and civil engineering. I could not stop thinking about the conversation the local people used to carry on about the villains who ran the mines.

I remember when the folks would come into the store arguing about and criticizing the mine owners. They would talk about the greed involved in the everyday operation as the miners and the shaft foremen were told to take large portions of the coal pillars. These were the things my father knew about and wasn't allowed to tell anybody. The pillars were supposed to remain in tact to support the surface above. The way the owners looked at it, there was a lot of coal in each one of those pillars. That was like leaving money behind. The loyal workers did what they were told to do. They were told to remove more coal out of the shaft than the specifications would allow. The veins were cut differently than they were supposed to. The specifications indicated one direction, but the mine bosses told them to follow the richest coal route. They never made the changes on the maps. But my father made his own marks to indicate the changes. The mine inspectors were getting paid to look the other way.

The economics of the region was stimulated with newly acquired entertainment and business ventures. Prostitution, gambling and the beer gardens, as they were called then, were evidence of prosperity from the beginning. The taverns were filled with laughter and joy. It was a poor mans' dream. The coal mining industry in the region produced wealth at a rapid pace. With the millions of men serving in the armed forces, unemployment declined. Demand for coal was so high that the government offered miners draft deference to stabilize the labor force and the miners worked a six-day week. The mineworkers now had far steadier incomes than before the war.

We all saw the good times. I remember going to Yarirsh's with my dad. Yarrish's was the local beer garden at the top of Minerhill. He would play shuffleboard for hours. From what I remember he was one of the best shooters in the competition. He also played the accordion and in between matches he would play polkas and ballads and everyone would join in singing to the tunes. The doors were open and everyone could hear it for blocks away.

My father told me that he bought the accordion for twenty dollars about ten years before and it always sounded like new. It seemed everybody knew and loved him. A lot of them spoke Polish and they wanted to sing Polish songs. They called us greenies in a most familiar and friendly tone.

After growing up in the neighborhood I got to know a lot of people. By hanging around with my dad I was getting to know as many of the older folks as well as the younger ones.

I remember how everyone had so much in common as they talked about the mines and every other business that was developing. The beer gardens in Minerhill were a home away from home for most of the people in the area.

The beer garden owners were compassionate to the locals. They ran a tab for many of the regulars, and on Fridays they were sure to have the bill paid. Friday was always the busiest time in the bars. Most of the patrons worked in the mines in one way or another, some in support, like truckers, and those who worked on the railroad. Many of them usually lived within walking distance and at times had a difficult journey home.

On Sunday mornings I would go with my dad to Rozaies, another local pub where he would meet his buddies. There, the gents would shoot pool and play cards. A lot of them, as they left their homes, told their wives they were going to church. More than once my dad got so drunk that on the way home he would hit the utility pole as he was pulling into the driveway at the house. My nephew Jackie always kept a spare left fender on hand or knew where we could get another one in a hurry. It was a regular occurrence.

It was still a great experience to be able to hang around and listen to what the elder people had to say. I was always listening to them and they seemed to have all the answers. They seemed to know everything. They weren't happy unless they were arguing.

My father died in 1968 of asthma or "black lung" as it was known in the mining industry, and a few years later my mother passed away. My dads last words to me were, "Make me proud, son, remember you can do it all."

After the military service I continued my education. I graduated from Pennsylvania School of Mining, and became a degreed civil and mining engineer. Unlike most my age, I stayed in the area where I was born and raised. I was able to get full time employment as a civil engineer with a local engineering firm, and I stayed with them for a long time. I performed building site surveys and sub- surface analysis, involving mine subsidence throughout the region.

As part of my job I also had to analyze risks associated with natural disasters, including wind, earthquake, fire, floods, and design structures. I had to deal with the safe, economic, and environmentally responsible extraction, processing and marketing of natural mineral resources from the earth, and organize the restoration, refilling and replanting of the land. I had the responsibility in this area to ensure that as little damage as possible was done to the environment. It was definitely a career that I enjoyed.

As a young educated engineer I now understood what the unions were fighting for. I now understood what the Molly Maquires stood for in their quest for a better working environment for the mineworkers. Now I understood more of the struggles.

The life experience I received growing up in a mining family and my engineering education made me aware of the corruption that took place in the industry since the beginning.

Chapter VI

The Later Years

It was now the 70's. I was forty-three years old. I married my wife, Susie when I was 28, and we now had two sons, Barry and Adam, and one daughter Taryn.

With the coal industry gone, the regions' economy is threatened. Most coal families have educated their children and sent them to areas of more promise. Very few companies in the area profiled any chance of survival for the blue-collar worker. With the coal went a great deal of other industries. Manufacturing, coal trucking companies, railroading, truck sales, machine shops, was now becoming a dim shadow of dreams. Buildings were being abandoned faster than one could imagine. The coal industry took its last breath just as the men who built it did. Those that remained from the coal mining industry still young enough to find employment were moving out in droves. Many of the older miners that were still alive were plagued with asthma and black lung.

A New Venture

Now a new source of corruption found its way into the valley between the coal dumps. Thousands of acres of so-called useless unclaimed land helped convince politicians and businessmen that this would be the place to take on new ventures and bring into the area what no one else wants.

The local chamber of commerce and the politicians were constantly making attempts to lure big business into the area. Very little blue-collar activity was taking place, but they continued to pull in some office jobs with insurance companies and financial groups. Manufacturing was sliding away rapidly.

They continued to convince each other that the area could withstand yet another defacing of the rich land that is placed under our feet. The plan was to remove, fill, or displace the land that had no significant real estate value and replace it with imported household, commercial and community's refuse. A potential major catastrophe is forming into reality.

Garbage in bags and cardboard boxes remain on the ground sometimes set on fire starting the coal underneath causing mine fires

They believed this could possibly be a way to help fill the mines that were stripped of billions of tons of coal over the last one hundred years. It could possibly make it useable once again for community recreation or construction in the years to come.

Spearheading its way into the future, this venture brought with it new spin-offs of greed and deceit. The importation of community and household garbage from other states gave temporary promise to residents that there would be a reduction in their property taxes if they would permit it. The residents in towns within the vicinity of the landfills jumped at the opportunity to have lesser taxes for a few years. The authorities promised that it would not affect the quality of life within their communities, and that the land could be used once again for municipal and community functions or possibly building projects. They soon found out it wasn't worth the inconveniences thrust upon them. The roadways throughout our towns were becoming burdened with heavy truck traffic, leaving mounds of refuse in their path. Many local roads were torn up by the weight of the trucks, and some of the contractors were making hauls with vehicles that were unsafe with bad brakes and tires. Injuries and deaths occurred frequently.

Landfills under scrutiny

Ample evidence suggests that organized crime controls much of the landfill industry in every state. All criminal activity is not entirely confined to organized crime; there is also evidence of other unscrupulous entrepreneurs and public officials, none of which would hesitate to jeopardize public health and safety. The extent of criminal activity within the waste disposal industry goes beyond corrupting the government. It directly impacts the best interest or lack there-of throughout society.

Organized crime was corrupting law enforcement officials and public officials through bribes, payoffs, and threats. It flowed out to dishonest businessmen who were making huge sums of money by cutting deals, breaking laws, violating regulations, polluting the air and water on a massive scale, paying off politicians, and violating health and safety laws.

Our environment and society continued to be threatened with an endless series of waste, leaving behind a legacy of damage.

The area had been inundated with problems caused by those individuals and companies, responsible for continued irresponsible activities. Those activities not only caused harm to the people in the communities, they also devastated the land. The severe punishment cast upon the lawbreakers was barely enough to impact the influx of illegal activity as it continued everyday. I couldn't help but to witness the destruction of the natural landscape of such a wonderful habitat.

Garbage Left To Burn

Chapter VII

The Territory

The years have gone by so fast. I am in my 70's. My children all live in Philadelphia, with their children. I am very proud of each one of them. They have achieved their goals and are dynamic adults. My son Barry is a financial advisor for a stock brokerage firm in Philadelphia. Taryn is teaching in Philadelphia and Adam is a corporate executive with a large engineering firm also in Philadelphia.

I've been doing pretty much of the same work as before I retired as consultant in the engineering field.

One of my jobs is monitoring mine fires. Mine fires create some of the most serious coal mining related problems. The anthracite area has suffered mine fires for centuries. I work on monitoring many of these fires in hopes of keeping the fire small and minimizing their effect on the environment. Many times we cannot put the fire out. In some towns, underground mine fires have been burning for years. We drill holes on these properties to monitor and work to extinguish the fire. The residents worry about losing their homes to the smoldering danger and inhaling the ash-based substance that is being used to plug openings to the fire. A lid is placed over the plugged borehole. When you flip the lid off, the heat, smoke, and stink come out of the mine fire hole. The sight smells like rotten eggs. The residents fear the fire will cause subsidence that will shift natural gas lines, causing an explosion, and worry about the deadly gases invading their homes.

Typically there are only few homes affected directly by these fires. Invisible from the surface, mine fires are tricky, as they travel through tunnels carved in coal seams decades ago. They're persistent and unpredictable.

Many people dump trash and other debris down in the old stripping pits. In time someone will torch the trash and some of these fires eventually reach the coal seams that are located at the bottom of the pits. Also, lightening hits old dead trees, and if the trees are located above an outcrop of coal, these coal veins will ignite.

In addition to the danger of the fire spreading to the homes and wooded areas, subsidence can occur when the blaze consumes the thick coal pillars left to help support the tunnels.

Sometimes we can excavate acres of land and try to extinguish part of the fire, by installing an underground clay barrier between the remaining fire and the community houses. However, years later the fire still burns around the barrier and needs to be diverted again. The goal is to further encircle the fire, entrapping it. The only way to guarantee the fire will be extinguished is to strip mine around the perimeter of the fire.

Coalmines were built with wooden beams to prop up their tunnels Thick walls of coal held the beams in place. An electrical storm, extreme heat, or even an above ground fire near an open coalmine vent can spark a fire down below. When the wooden frame of a mine catches fire, the surrounding coal provides enough fuel to keep the fire burning for years.

A fire still rages in Colorado. Its temperature rises higher than 1000 degrees Fahrenheit. It has burned out of control for more than 50 years. This fire is monstrous, but no one has ever seen it, because it is blazing 65 feet below the ground, in an old coalmine. The oldest known continuously burning underground mine fire is in New South Whales, Australia. They think the fire started when lightening struck an outcrop of coal near the surface of the ground.

In Panther Valley, Pennsylvania, a mine fire that became known as "The Burning Mines," became a famous tourist attraction. At night, the view of the fire that broke through the surface was awesome. Tourists came from all over the world to view "The

Burning Mines." Eventually the owners dug a huge trench across the mammoth vein of coal that was burning and placed a concrete and clay barrier to stop the fire. These helped somewhat, but other measures of underground dams and water eventually had to be used.

It is nearly impossible to fight a fire that you cannot see, so you have to find the hot spot. We can't see what we are looking for, so we drill holes in specific areas based on old mine maps. We drop a gas-detecting meter down each hole drilled. If carbon monoxide, sulfur dioxide, and methane are present there's a fire burning nearby. We also drop a thermometer in the hole to check the underground temperature. Water cannot be used to stop the fire, because water boils suddenly and with nowhere to go, an explosion would occur.

Once underground mining shut down, any inaccessible mine fire starting underground was left to burn. And the town would be evacuated as the underground coal burned profusely. The dangers of gas line explosions, sulfur toxins, fire, and subsidence forced hundreds of residents to flee the vicinity. On any given day any reaction to the coalmines could cause a disaster. The town would become a ghost town.

Most of the old miners that worked the mines are deceased and a few of the younger residents of the area don't have the slightest idea of how far the mines traveled through the valley. It would be difficult to give precise mileage of the mines since they serpentine in so many directions.

Once in a while I would pull out some of the old maps I had and give a demonstration of the mine layout throughout the valley.

In fact, I was asked to speak at a local Rotary meeting on the mine memorabilia I had, including the maps. There are different maps and memorabilia on display at many of the local historical sites, but none that tell the story like the ones I had. After all, my father marked all of the maps of the underground tunnels of the valley as they really were. He traveled the entire valley in and out of the mines. The maps I had were more accurate maps than any others. These maps were even more accurate than the maps the bureau of mining composed.

Many times I tried to explain to various individuals that there was more subsurface material taken from beneath than anyone could ever imagine. I showed copies of some of the original maps that the United States Department of Mines had in their possession, and I matched them with the ones I had. Nobody could accurately calculate the difference in the tonnage of coal that was removed compared to what was originally claimed.

Stephen A. Wzorek

Newspaper cartoon critical of anthracite mining practices. (Courtesy of Sunday Independent)

The maps were vastly different. The coal barons robbed pillars and ceilings; instead of following the contour of the land, they followed the coal. `

I remembered the Port Griffith disaster, and the similarity: "It was January 22nd, 1959. The miners were instructed to leave a structure less than safe between the river basin and the mineshaft. At one point where the shaft traveled under the Susquehanna River the ragging water trapped seventy men when the ceiling caved in. The disaster took twelve lives with it, all because they were told to take more coal than engineering would allow.

Stephen A. Wzorek

Anxious families of miners during rescue efforts after the Knox Mine Disaster. (From the book Knox Mine Disaster)

Again, the miners were told during the operation to take the coal that was to remain as structural support pillars. The extra tonnage of coal that was constantly being removed illegally and the total of criminal acts are still continuing. This is just more evidence of the cruelty to the worker.

Coal mining is almost non-existent in our area, but the remains are evident. The coal industry fatally ended with the Knox incident. It was national news .

It was always an eye opening experience for the young people to realize the miles and miles of underground mines that stretched through the valley. A different company owned each field but they were all tied together. Much of the mines, which are tied together presently, cover as much as 200 lineal miles through the valley, much of which is partially filled with storm runoff water. In fact, there is so much water underground it could surely be categorized as one of the largest underground rivers in the country.

One particular vein spread through the lower valley and is known to be crossing the river somewhere in the east central part of the state in the vicinity of Dalton, and Fernwick. These are places on many occasions that have hot spots or burning underground veins.

As new buildings were going up, the company I worked for performed core-drilling samples to see what the subsurface had in store. A majority of the time we would find solid rock, but no one was ever surprised when there were voids under the immediate area. It usually was a matter of how far below it was and whether it would sustain the weight of a large structure, which was an everyday concern.

The evidence of shallow mines in the area and the subsequent subsidence they cause kept me busy. The coal pillars that supported the ceilings of the mines kept them from collapsing but had weakened by years.

As mines age, supporting coal pillars decompose and eventually can't hold the weight above the earth. Deterioration of the roof of the mine and subsequent surface subsidence takes place over many years. Just about every town around the valley was deep mined.

People can live in their homes for years without knowing the mine is underneath them and about to collapse. Many people begin to notice that their doors don't quite shut all the way. Others notice small cracks in the walls, ceilings, or foundations. The foundations sometimes separate so much that rain pours into the basement. Eventually the foundation starts to slope so much that it damages underground water tanks or gas lines causing dangerous leaks. People can hear their houses creak and watch the house actually sink, causing major structural damage.

In the deep mines, pillars of coal laid out in a path formation, were supposed to be left in the mine to support the land on top of it. Many houses are at least partially located on an underground block of coal but empty spaces where the ground was mined can still cause problems. If rotting timbers or a shift in the earth in one section of the mine occurs, it can affect another section as well.

It was my job to design the specifications for stabilization, advertise for bids for stabilization work, and recommend the contractor. We would stabilize the underground areas by several techniques. The most common method is bulk filling of the void. We first stabilize the area by boring holes down to the mine void and injecting a cement mixture. First, holes, three to six inches in diameter are drilled 120 feet into the abandoned coalmine and then they are filled with grout to stabilize the land.

Another method is constructing columns of grout in the mine. In areas where the mine is open, grout is pumped into the mine. As the columns build, the injection pipe is withdrawn until it is even with the ceiling. The amount of grout injected is determined by the height of the mine roof.

Once the stabilization is done and the damages are repaired, there shouldn't be any more problems. Some houses can be moved back and the foundation fixed to correct the problem after the ground has been stabilized.

Chapter VIII

The Coal and the Garbage

Over the last ten to twelve years trucks have been hauling garbage to the areas in the Anthracite Valley at as much as 200 tons per day.

The haulers and landfill owners are constantly paying fines for overloaded carriers and other various incidents. In the past, there have been criminal charges of accepting illegal waste. Waste such as fluff and chemicals were found to be filtering into a compound that was only licensed for household solid waste. Since then the landfills are monitored more closely.

The solid waste has come for over ten years and no one can stop it. Politicians promise to limit the tonnage and curtail the expansion privileges demanded by the landfill owners. But that's all it is, a promise.

In the meantime the solid waste is mounting 100 to 150 feet into the sky on hundreds of acres. All that can be seen is the partial black membrane and small vent stacks proportionally spaced and covering the composting waste.

As the waste starts to decompose, new techniques are being implemented. The liquid remains that drain off the garbage are being harnessed by plumbing hidden beneath the plastic membrane. As the liquid or, leachate seeps to its lowest level, it is directed to collectors for treatment and is headed, eventually, to city sewerage systems and secondary streams and rivers.

Once the decomposition has completed its cycle, a section of the landfill is then capped off. Residents all over the county have been trying to stop the influx of illegal landfill deals for years. The bullies of the garbage industry are overriding the NIMBY (not in my back yard) syndrome. The residents don't stand a chance.

The landfills in spite of all the public oppositions are growing at an enormous rate, covering a large amount of land on both sides of the valley, just like the culm dumps. To this day they've grown to a point where enough methane gas is being produced to generate sufficient electric power to sell back to the utility companies at an enormous profit. Although there are regulations in place governing the landfills, there is not enough personnel to police the daily activity that takes place. While most of the landfill owners are operating within the perimeters of the law there are many that take shortcuts to maintain the profits. It is at this particular time that the danger becomes eminent.

Chapter IX

Our Town

The production of incoming garbage is far surpassing the decomposition rate of landfills, and certain procedures have been ignored over the years.

Meanwhile, in March of 2000, a major highway project is taking place a mile from the town of Dushore. The highway is a major hub from Canada and New York City and other points on the East Coast. The highway department and contractors are dynamite blasting almost daily. The local hotel complex has filed complaints of damage to windows, and of noise and dust from the blasting. They and their near neighbors have been subjected to never-ending annoyances either from thick, fine dust that covers furniture and cars, or from the noise of dynamite blasting, which shakes their dwellings.

The housing development on the other side of the highway has residents complaining of a smell of a natural gas odor. They feel the blasting causes it, and they are certain that the blasting has erupted a natural gas line in the area.

The utility companies were called in to look for loss of line pressure or evidence of natural gas leaks.

The main component of natural gas is methane. It is lighter than air. So if there are any leaks it does not accumulate around the "source" of the leak. The gas company injects a chemical called ethyl mercaptin into the gas flow of the pipeline and distribution systems. They do this as a safety precaution, to give gas an identifiable odor, since natural gas is odorless and tasteless in its' natural state. The odor is compared to the smell of rotten eggs or garbage. This odor is noxious to get attention and action.

In the early morning hours of March 10th, I was eating my breakfast, when I got a call on my personal pager. There had been an explosion. Gas seeped into the basement of Janine and Steve Starinsky, who reside in the development, causing the explosion.

I went to the scene to see if I could be of any help. I was horror-struck by the sight. The house was to the ground in splinters. Fortunately, both parents and their two young sons were out following their usual work and school routines. Neighbors said they heard a thunderous noise like a plane crashing. Rocketing debris flew around the neighborhood like a tornado hit. A slab of roof landed in the next-door neighbors yard. Flying glass embedded itself in a tree on the other neighbor's lawn, with glass protruding out of the tree. This same tree held a flying portion of a picture window. The flattened house burned and smoldered for a few hours as firefighters fought the blaze.

Three local fire companies with five trucks, ambulances, two battalion chiefs, and the hazardous materials crew responded to the three-alarm emergency and set up a command post near the development. By afternoon, the rumors were abuzz on television and from other news media. Many residents left their homes because of the gas odor. Eyewitnesses were beginning to talk about what they experienced after the highway project started dynamiting at the beginning.

The Starinsky family had unusually high levels of the poisonous gas around their home. Inspectors checked everything for some kind of clue: the blackened stove, mangled heaters, and anything else that might give off carbon monoxide.

They were ready to give up on locating the source when an inspector realized that the carbon monoxide could have come from the dynamite blasting several blocks away where the highway project was taking place. The dynamite could have released the gas underground, and the gas might have formed a bubble under the family's house.

Carbon monoxide readings were taken in the houses in the development. Methane gas and nitrogen oxide levels were also monitored in the houses. Soil samples were collected around the houses and tested for carbon residues. The investigation revealed the presence of pockets of carbon monoxide under the foundations of the houses. The investigation revealed the probable source of contamination was the use of explosives at the highway construction but where were the gases coming from?

Warnings and what to do if an odor of gas was detected interrupted television programs. Peoples Gas Company spokesman, Tom Comerford, in a calmly manner, cautioned residents with his statements, alerting them to be aware of the gas. He said, "If the odor is strong, leave your home immediately. Notify the authorities by phoning the twenty-four hour hotline at 1-800-GAS- LEAK. Do not use a phone inside the building or near the gas leak. Also, do not operate any light switches, or electrical devices. Any of this activity could ignite the accumulated gases." He went on to say: "I want to alert the entire town that the main gas line has been shut down to all of Dushore Development as a precautionary measurement. The police have barricaded the roads into the development."

Large forced air fans were set up about sixty feet into sewer pipes to blow the gas out through a manhole inside the neighborhood.

Luckily the homes were not too close together. Surrounding buildings around the Starinsky home sustained only a minimum amount of damage from the explosion, and a hazardous material team was surveying a one-block area for signs of additional gas leaks.

You could smell the odor of gas in the air. The emergency escalates as immediate area residents are told to evacuate, starting from one house to another, to twenty-two homes in the small Dushore Development.

The emergency response coordinator, fire marshal, Ed Connors, is heading the investigation, along with Tom Bullaro, a public utilities engineering specialist, who investigates gas explosions.

I spoke with Ed Connors. He stated: "This kind of explosion is uncommon. I have heard of only about four like this in the last twenty-five years. Gas leaks are relatively frequent occurrences, but this type is not." As he shook his head he said, "It takes a rare combination of circumstances to cause disruption like this. Gas leaks aren't very hazardous except when they find an avenue into a home and then have enough time to build a volume up to a dangerous level. Gas has to be escaping from some large source and finding its way to this development." Connors cautioned: "The escaping gas can either rise to the earth's surface and escape, or if it can't reach the surface it could migrate under ground until it finds an escape path it could follow." Connors concluded, "I think we have a larger and more dangerous situation here than we had imagined. Since we don't know where the gas is coming from, how do we know where and how far it will travel underground? It could eventually take out everything in its path."

My only thought at the time was how much time did we loose and how much time do we have before we had a catastrophe.

Tom Bullaro said, somehow the gas got into the house, built up and ignited. Tom relayed to us, "Gas goes upward and will blow a roof off, as it did in this case. In most cases a third party causes the damage, but there is no evidence of any construction activity here, nor is their explanation for where the gas came from." Tom also added, "Even with a gas leak, there's a very small window of opportunity for an explosion." Since we had no idea how or where the gas was coming from or where it was going to, there was very little comfort in those facts.

Gas crews continued to drill ventilating wells they hoped would relieve the gas bubbles that had developed under the development.

Meanwhile the gas utility workers walked and drove the streets of surrounding communities testing for irregular gas levels. They tested the air in every nearby community, trying to determine if any and just how much gas was under the city. For some unknown reason, there seemed suddenly to be much higher levels in every community tested. It was illuminating a dangerous problem, and horrifying the residents, who town by town were told they too must evacuate their homes.

The local Gas Utility Company was already getting calls from residents from miles and miles away. The utility companies were called to inspect the troubles as the calls came in and they indeed detected the gas. But still, they couldn't locate the source.

While investigators on the scene were becoming increasingly alarmed, the city officials were on the local newscast, downplaying the devastating turn of events, for the benefit of the public.

CHAPTER X

THE DANGER ESCALADES

Bill Brezevoy, a civil engineer from the highway project arrived on the scene. He suggested calling someone from the Bureau of Mining and Engineering. He implied suspicion that it could be the blasting disturbing the mine gases.

Ed called the Bureau of Mining. He spoke with mining engineer Bob Webb, who told him, "There is no indication of any mines in that area according to our maps. The maps show the immediate geographical area does not indicate any shafts or mining activity in that area, I suggest you follow Mr. Wzoreks intuitions." Bob added, "Ed, Steve has the experience and the knowledge of the local mining industry."

Bob didn't say it, but he knew that I had the different maps in my possession. I started to form my own opinion even before I received the call. But to my own disbelief I couldn't imagine the conclusion that was inevitable.

THE EVENT

Day one

We decided to set up a second command post at the American Legion Hall in Dushore. The legion hall was equipped to monitor all reports coming in for possible gas leaks, smells, or any related problems.

After looking over the gas companies' records of the past weeks, we saw no indication of a natural gas line break. Everyone was convinced it had to be coal gas, which they were getting readings from. They showed me their maps and I had a photographic de's je vu.

We studied maps from the archives of the bureau of mining and engineering. I remembered seeing this particular layout before. I said, "Look, I know the maps you have are official, but I have to tell you these maps to my knowledge are not accurate. If you will excuse me I have to leave for a while and I will be back with some better information for you."

I quickly went home to the basement of my house, and I pulled out a few old boxes, before I found what I remembered seeing before. The map that they had displayed at the legion was similar to

111

one of the maps that my father had given me. One that I remember Marty and I playing with . I unfolded the map and there it was, the penciled lines that my father sketched over the original survey lines. The original lines of elevation and geographical location were actually lighter than the ones my father sketched on the print. The estimated amount of coal that was to be extracted from the area was written next to the designated area. My father also had an estimated amount penciled in.

I took my map to the command post, and we started laying out the direction of the underground caverns that followed the coal vein, as the map showed. We worked for two days on this. By the third day it took us back to somewhere at least two miles away to beneath the landfill on the other side of the highway. By this time we were showing concern for the possibilities. But no one really wanted to admit to anything just yet.

My maps showed masses of excavation beneath the entire area of the landfill. The original ones didn't.

One shaft showed up directly below one of the original landfill heaps. This site was capped off and abandoned at least ten years ago according to the landfill operator. Since then the rules and regulations have changed for landfill operators, and in recent years laws were passed requiring landfill operators to perform core drilling samples and on-site tests by the EPA, before it could be capped. This area was obviously not held to those specifications.

I said to myself, "Is it possible that someone ignored the rules and someone might already know something that we don't?"

I finally had the answer, "The huge decomposing heaps that were capped were still active and the gases were being evacuated without control and no one knew it for all those years." Tom Bullaro asked, "But where was the methane going?" I unbelievingly said, "There were no monitoring devices and no way to tell how much gas had been evacuated, or where it went."

It was March 19th, Tuesday, 10:00 AM: we were assembled in the hall getting ready to compile all our findings and take the next steps to follow our problem, when the call came in. A gas explosion had just occurred at a residence in the Minooka area, just about 6 miles from the Legion.

All of a sudden all eyes and ears were on me. They wanted me to make the next suggestion on what had to be done.

I suggested, "The first thing we need to do is to go to this explosion to find the cause before we jump to any conclusions."

ED Connors and I went to the scene. By the time we arrived, the fire was extinguished. A small wooden framed home was partially demolished. One wall was leaning out at an angle, ready to collapse. After a thorough inspection, Ed concluded, "The explosion was caused by gas leaking into the basement. The accumulation of gas found the furnaces ignition, and triggered the explosion. I sighed, "We have to stop this flowing gas fast. We were very fortunate this time that no one was in the house, and there were no injuries or fatalities."

CHAPTER XI

There was no doubt that the evidence pointed to methane gas. We knew it was highly explosive, but there was no way I knew the specifics.

I contacted my brother-in-law, Bruce Anton, who lived in New Jersey. He is an engineer for a chemical company. I knew he had performed consulting work for large energy resource companies over the years. I thought maybe, if he didn't know the answers he would know someone who did.

"Bruce, "I have a situation here and it's getting out of hand. We have a severe flow of methane coming from the abandoned landfill and it has been filling up the mine tunnels and shafts. At the moment we have traces of the methane showing up miles away from the landfill. We need to know the specifics of the methane and is there any way to capture it." I told him: "Get back to me as soon as possible."

The next step was to locate the areas of the regions that were complaining of gas odors and to chart them out. The finished chart revealed a path that was leading away from the enormous landfill on the east side of the valley. There was no time to waste.

Bruce got back to me within the hour with his findings: "It can move away from the landfill site through cracks in the earth or bedrock. The movement of the gas is affected by weather conditions. It can only move gradually upward along natural pathways or along manmade paths such as pipelines, or in this case mine tunnels. The only safe way to extinguish a flammable methane gas fire is to stop the flow of gas. If the flow cannot be stopped, cool the surroundings with water from a suitable distance. Extinguishing the fire without stopping the flow of gas may permit the formation of ignitable or explosive mixtures with air. These mixtures may spread to a source of ignition. It has a tendency to associate with cool moist air, traveling through natural or manmade pathways, and that makes it an ideal candidate to follow the mine tunnels.

I asked Ed Connors to call a meeting through the mayor's office, inviting representatives from the local utility companies, the engineers from the highway construction project, the landfill and state representatives. Many of the news media got wind of it and they were already there.

We met at the legion and I discussed the possibilities. We had all sorts of opinions as to what could be happening. We were monitoring the calls that were coming in for the odor of the gas. As we watched and marked the map we could see it taking a direction south of the landfill, right down the valley as far as twenty-five miles away.

As we were discussing the possibilities everyone had different theories. After hours of discussion we concluded that the landfill had never completely stopped manufacturing the methane for all the years. It just stopped for a while, and then while capped off, it forced its way to least resistance, down through the mines.

The blasting that took place miles away during the highway construction over the past several months opened crevices to the illegal mine tunnels and as more blasting occurred the crevices opened even more.

It just seemed like there couldn't possibly be that much gas in one landfill. The owner of the landfill was at the meeting. He admitted, "We did notice a severe drop in the amount of methane that was expelling from the flutes. It seemed to have dropped since about a month or more, which was unusual." This was about the time that the highway blasting started. The dynamite blasting combined with the amount of coal illegally removed from the areas subsurface and the illegal construction of the landfills that were left abandoned has created the potential catastrophe.

It was getting more dangerous than we thought. It seemed more evident that we were now dealing with two sources of the methane gas. If this were so, could they join together anywhere in the valley. Tom Bullaro questioned, "How much capacity could we be dealing with?"

I told everyone at the gathering that I had to return home to locate some more of the maps. I wanted to see if they showed the route that the gas could possibly follow from the West Mountain.

Hurriedly, I ran to my car, turned the radio on to listen to anything but more bad news. When I got home, I saw my sons were there, trying to calm my wife's fears. They were obviously concerned as they watched the developing news on television. As soon as I got in the door they started asking questions. "Are they going to evacuate everyone? Should we leave the area?" "Are we in any danger staying here?" I couldn't answer them fast enough. Angrily, I shouted, "Everyone is asking me questions that I can't answer." Spotting my favorite chair, I flopped in it and took a long sigh. "I'm sorry for the outburst. It's just that I'm so tired and frustrated. I don't know what's going to happen, but I am doing my best to stop whatever it is. This whole thing is getting out of hand. It wasn't supposed to get to the public this way." The neighbors saw me on the news and now they were pounding on my door to find out what to do. I tried to calm and assure them that everything was being done to get to the bottom of the problem. But they wanted some solid answers and I didn't have them. I was wasting time and needed to find those maps. I went down to the basement, gathered the rest of my maps and flew out the door.

I headed back to the meeting place at the legion. I rummaged through the boxes for the map I needed. Sure enough it was showing the veins of the mineshaft winding along the west side of the Lackawanna and Susquehanna Rivers. More lines penciled in by my dad.

Calls were still coming in of flooding and underground explosions, just below West Pitston. That was at least twenty-five miles down valley on the west side of the Susquhana River. This was where the two rivers met. It was where some of the deepest mine shafts were as deep as 250 feet below the surface. Some were known to cross under the rivers. It was near the area where the Knox Mine disaster occurred.

We started looking at the Upper Valley maps when Mr. George Kubasko, from United Environmental arrived. It was about 4:00 PM, and we began to exchange information.

I told him that we knew that the temperature could be as low as 40 degrees Fahrenheit in the deeper shafts. He told us, "It wouldn't be difficult to imagine that the methane would be like a serpent looking for the cold wet air. The colder it got the more that the methane will accumulate and consolidate." We spent the next three or four hours just trying to analyze the situation and assess the possibilities. We knew we had to take steps to protect the population, but we didn't want to cause a panic or force a mass exodus of the entire region.

CHAPTER XII

More Deceit

Representatives from the different utility companies, the mayor's office, and other bystanders patiently waited for some type of decision.

As I was looking over the pencil-scribbled maps, I couldn't stop thinking of my dad. I started remembering the stories told by the people in the old store. I thought of the stories that passed back and forth as I listened to them talking about the mistreatment and deceit by the mine owners.

I said ironically, "We now have the spin-off in dealing in the mishandling of the refuse industry. The greed is haunting the entire community." Criminal acts were traded for profits once again. The profiteers were nowhere to be found. The refuge industry covered up their illegal business strategies, as the mine owners covered up their wrongdoings. The only ones hurt were the workingman. We cannot locate one individual to hold responsible. Litigation would take years. We had to act fast to the wrongdoing and stop the pain.

I knew a few things for certain. We had a very serious disaster impending. We were dealing with methane gas that liked cool moist air, it was flowing south filling the mines, and third, we had two sources and it could explode at any time with the slightest ignition.

I laid the two maps on the table next to each other. They showed the shafts were parallel on both sides of the river for most of the route. The shaft on the west side seized just on the other side of Burwick.

Just at that time someone made the statement that the nuclear power plant was in the town of Burwick. Luckily the vein on the west side did not precede that far.

I glanced at the map showing the east vain. The vein snaked its way along the Susquehanna River. It crossed under the river directly toward the power plant. It pointed right at the remains of the old #8 breaker, next to the power plant. It was obvious as everybody remembered the mountains of culm that still remain in the specific area. This too was one of the deepest shafts in the southern part of the valley.

If the methane were to follow this path there would be not only a gas explosion, but also fallout from the nuclear power plant. It could spread hundreds of miles throughout the northeast. I was extremely distressed now, "How are we ever going to stop this from happening?"

Just as we were discussing the results of the findings, the earth vibrated below us. We heard what sounded like thunder beneath the ground. "I held my breath, "Oh no, don't let this be happening!"

We asked to be informed of the disturbance as soon as they found anything out. A few minutes later there was an announcement that there was an explosion near Nantecoke. The only available information the dispatcher had was, "Several people have been rushed to hospitals with various wounds, after being hit by flying debris, as they were driving down a street. Two of them are in serious condition." We were told it didn't reach the power plant.

We had to do something immediately. I inspected the map and the lines my father drew. He had elevations and the volume of the veins hand written with pencil. Next to them the vein that was taking its course under the river was far below the river. It was deep and wide. There must have been a lot of coal in that vein.

Ed Connors, Tom Bullaro, Bill Brezevoy, and I, decided to go to the river in the vicinity of Burwick. It took us about thirty minutes in the SUV to drive down through I 81 south and over the Cross Valley Expressway toward Burwick. From the distance as we traveled down the expressway we could see the smoke surrounding the Nantecoke area.

When we arrived at the river site in Burwick, we decided to investigate the latest blast. We followed the smoke, which led us to the area of the old #8 breaker.

The day had been intensely hot and humid with thunderstorms rumbling through the valley. A severe thunderstorm with gusty wind forces and lightening hit the #8 shaft. When we arrived, the thunder was still deafening and the rain was a torrential downpour.

The explosion must have developed intense force, as the blast ripped out the pavement on the streets over a four-block area. Cars and trucks were hurled up against each other like dominos, or they were crushed by large slabs of concrete. It looked like a war zone.

Rescue workers were feverishly using chain saws, hacksaws, and crowbars to get to the trapped motorists.

The force of the explosion blasted up some large rocks and shale, some weighing up to 1000 lbs. and filled about a hundred feet of the nearby Nantecoke Creek bed with shattered stone.

Trees were uprooted and toppled over and the creek was partly blocked, backing it up. The water level up stream was about three feet higher than the lower level of the creek, even though the water was seeping underneath the rocks. There seemed to be a huge hole under the creek. You could hear water running and gas was bubbling up all around the banks. The water shot up out of the creek like a gusher, where some large rocks had been trapped.

Large pieces of shale and rock landed about ten feet from the creek bed and mud was blasted out of the creek from the explosion.

The streets that were affected by the explosion followed the course of the storm sewers. Ed thought aloud, "With the ground so saturated and the force of the blast, could the methane gas seek higher levels of moisture?" Nobody knew the answer.

We met up with the Nantecoke fire marshal and other inspectors who were just beginning to put together their findings. After a thorough investigation, they came to the conclusion that the lightening had hit the rails of the mines boxcars which were left abandoned and traveled from the surface down into the old shaft, igniting the pockets of methane and leftover coal dust. The #8 breaker was a gassy pit, having high concentrations of methane with its coal. This methane continued to bleed off into the mine workings years after being abandoned.

We became alarmed about the nearby wells and the sewer systems, since there were so much left-over-trapped gases probably in the old breaker. "I questioned, "Where would it all go? Would the chemicals and gases from the landfill come into contact with the mine gases through the soaked ground?"

All of a sudden the earth began to tremble. We soon started receiving multiple reports of water main breaks and flooding.

The reports were becoming alarmingly more frequent.

Reports of manhole covers flying off, and the pungent smell of gas in the sewer lines. The force of the explosion caused covers to sail through the air and crash into car windows and roofs.

Up and down the valley reports were coming in of underground explosions. Sidewalks were cracking, crumbling, or disappearing into opened holes. The ground was cracking and swallowing anything in its path. Some towns had utility poles toppling over; twisted steel, smashed glass, and rumble littered torn streets.

The accumulation of the chemicals from industrial waste seeping into the ground and the coal gases from the abandoned mine shaft accounted for the magnitude and widespread of damage.

Meanwhile we had a call from the owner of the landfill. Apparently a recycling employee detected a strange odor at the site of the now closed landfill in Dushore. After investigating, he found gas flowing up through the ground in the area of a steel pipeline that ran through the landfill. He immediately called the 1-800 Gas leak number hot lines and reported the gas on the ground.

What we feared the most was now happening. The ground was collapsing and sucking up or destroying anything in its path.

Large holes were left like old mines without their roofs, cold, dark, empty. Like the mine bosses before, now the mines were sucking life out of the land once again.

The only explanation of the rest of the area not blowing up from the explosion was that the heaviest part of the methane had collected in the deep mine at Nantecoke. Once the explosion occurred it caused an implosion, sucking in the gas from the rest of the vein. Then once it ignited it was just one large blast and out. The damage was done.

Apparently, and luckily, it didn't have an effect on the other vein on the east bank of the river. The maps indicated that the east vein was a lot larger and could carry a larger volume of methane. The landfill was also much larger. Therefore it wasn't moving as fast down valley. But the fact was the landfill on the East Mountain had been emitting a great deal more methane for a longer period of time.

When we finally arrived at the spot we located on the map where the vein crossed under the river, we found an old rusted culvert pipe shooting vertically out of the ground. It must have been installed as a vent, put in by the miners approximately fifty years ago. We put a gas detector, snuffer, over the vent. We took the readings and found no gas. We got in the van and drove over the bridge on Eighth Street. It took us over to the other side of the river to where the map indicated that the vein entered under the river. Another vent was located right at that point. This gave us another opportunity to take some tests. We took the readings and again, our readings were negative.

CHAPTER XIII

TECHNOLOGY TAKES OVER

I suggested that we contact Colonial Construction, the company that was working on the highway project. They had a fiber optic camera. It could scope the vent pipe and follow the vein by remote control. We could send it in to see how much water if any was in the vein under the river. According to the map the vein was well below the river. It was one of the few that wasn't scabbed. It wasn't penciled on the map. For some reason the coal barons didn't steal the extra coal.

I got on the cell phone and placed a call to Colonial Construction. I explained what we needed. I impatiently asked, "Can you come to the site as quickly as possible?" He must have heard the urgency in my voice. Within an hour they had their crew with us.

In the meantime we were tracking the calls on the gas odor complaints. It was difficult to tell which calls were associated with our source of the problem.

There was panic everywhere and some people assumed they smelled the gas. Some of it could have

129

been from the explosion, but we were not getting a pattern. I contacted the utility companies and asked them to track the calls as the complaints traveled south. I told them to meet us at the river site. They located the last call near Pitston, about twenty-miles up valley north of the Burwick Power Plant. I sent them to look for any vents near the old breakers anywhere near the area. Within an hour they returned with the answer to my request and found two different vents. It was a good break for us. One vent had the gas detected, and one just a few thousand yards south did not. According to the map it was matching the route we were following. I was sure the gas was somewhere in that territory and we had to stop the flow from reaching the power plant.

The Possibilities

My thoughts focused on the damage that could be potentially caused if the methane reached the nuclear power plant. The power plant employs about 1000 people.

Until we contact the regulatory commission there was no telling how much uranium is stored on the property. It was said that if the power plant ever malfunctioned, it could devastate a 50 to 100 mile radius of the plant with fallout. We had to divert or stop the methane soon.

I called Bruce Anton again and asked him if it would be worth trying to recover the methane. He in fact said, "It would not be a good idea to vent it or burn it to the atmosphere. Methane has a severe impact on the ozone layer. It has to be controlled."

I called the mayor to see if he had a contact at the Penn Landfill. He gave me Marc Rena, the chief operations engineer. I called Marc and asked him if there was any type of portable methane recovery equipment that we could get to the vents we found. I told him "We located vents just below where the methane was last located." Marc, I need this equipment at the sight quickly, so we can recover the methane before it gets to the power plant and causes a catastrophe." He said he would call SBM INC, a company that specializes in methane gas recovery projects. I told him to contact them and see if they could get some equipment out here immediately.

Quick Action

We all concluded that we had to take drastic action and try to stop the methane from crossing under the river. The optic scope revealed that there was enough opening in the vein for the gas to pass under the river to the power plant. I had an idea of what had to be done to block the methane from getting to the power plant. Once again I got on the cell phone.

I called Kevin Brooks from Brooks Construction and asked him to bring the biggest crane bucket machine he had. I knew from working with him on past projects that he had crane machines that would dig ten ton at a time.

We had to find out approximately how fast the methane was traveling so then we would know how much time we had.

Bill Brezevoy and Tom Bullaro agreed, we needed to calculate the movement of the methane.

We needed the measurements of the excavation in the area, including how far away it was and what the pressure flow rate was at the first vent. The P&J Utility Company could take the measurements that we needed. They immediately began taking the readings and they relayed them back to us.

As we were calculating the movement of the methane, Brooks Construction arrived with the big machine. I showed them where I wanted them to set up. They said it would take about a half an hour.

There was a lot of activity in the area and just then Mike Griffiths arrived with SBM and their recovery equipment. I showed them where the vent was that they could connect to. As the gas company was constantly taking readings in the area, they noticed they were just starting to get a very slight movement on the meter.

I then warned them they had less than three hours before the methane reached the area. We had to stop it from traveling under the river. This was our only chance.

SBM had the methane recovery unit connected and started carefully vacuuming the gas. As the vacuum pump pulled the gas, a unit afterwards compressed the gas and transferred it to large holding tanks behind the truck. More trucks with the storage units were ordered and they were on their way. Each truck would have to be moved in one after the other.

The bucket crane began digging next to the river's edge. I told the crew to dig until they hit the mineshaft. It had big cables on a 60-foot high crane that could extend out 40 to 50 feet. It was removing the earth at 10 tons every shovel full. It took about two hours to break through to the tunnel. You could see the glare off the water at the lowest level. This was our only chance to cut off the gas from the power plant. If the methane crossed under the river the electrical frequencies alone could set it off.

The Call

This was our only chance now to stop the methane landfill gases from coming together with the mine gases. The path that the methane might possibly follow was approximately 30 feet from the river. The vent pipe was approximately another 150 feet north of the hole. At the hole you could hear a small amount of water running into the shaft.

If we flood the shaft with the river water it will flow in until it seeks its own level. According to the drawings there is an elevation steep enough that once the water seeks its level it will block the shaft.

I ironically said," Who would of thought that the waters of this river that took twelve lives over 40 years ago would flood the mines again to save the entire region from a disaster."

I made the call. I yelled out, "Start digging from the hole to the river until the waters begins to run." It took the machine about 45 minutes before the water started gushing into the large void. It was whirl pooling with an immense force, until finally we heard a one-time sound, like the mine took its last gulp.

The pressure at the recovery equipment immediately started to build up. The crew from SBM assured us that it was controlled and there were no problems with keeping up with the flow of the methane.

"Ok, this is our last chance. We're either heroes or I'm a fool," I said. All eyes were on me. I could feel it as if they were actually touching me.

We were now at the critical stage. There were more and more people surrounding the area, but still, all you could hear was the rumbling of the river and the roar of the machine as it was still tearing into the riverbank. As we stood by watching and monitoring the events, we were convinced that we had prevented the methane from reaching the power plant. There was no telling how much of a disaster could have taken place if we were not able to pull this off. We achieved our goal.

The news media eventually moved closer to where we were working and were monitoring our communications as the events took place. They kept abreast of the situation as the gas flow pressure was constantly changing. Every major television station had a crew filming our every movement. There were newspaper media and radio announcers continuously talking to their representative groups.

While we were talking I could hear some of the comments that each newscaster was passing on to their stations. They too were nervously anticipating the final outcome that would put an end to this tragedy. There was so much damage already done. I stopped for a moment to ask if anyone of them heard anything on the incident at Nantecoke. They said there were quite a few injuries but luckily there were no fatalities. I sighed, "Thank God, I will be so relieved when this is over."

Ed Connors called over to the police to ask them to cordon off more of the area. We were getting closed in. After all this, we didn't want anyone to get injured by the heavy equipment or to fall into the excavation. The crowd was getting large. It seemed that everyone from the legion came to the scene.

Meanwhile, I was in constant contact with everyone operating the equipment. I kept a watchful eye on the gages as dials increased to show a heavier flow. There were at least 10 gages involved. We were waiting for everything to stabilize to a point where the pumps on the equipment were at a steady idle, just enough to maintain a regular flow. The gages all had to equalize and have the same reading on both sides of the pump. It took a few minutes and I declared, "It looks like we're getting close." Just then the motors on the pump went into a low speed idle and the gages all were reading the same pressure. The methane was now under control. I shouted, "We have containment."

The somber crowd now clapped and cheered congratulations.

As I looked around I saw my wife Sue, Barry, Taryn, Adam, and my three grandkids.

I told the police to let them through the crowd. They started running toward me, and I could feel a rush through my body, but this time it was a good, comfortable and warm rush. We all wrapped our arms around each other, when I heard my fathers voice, "Stevie, you made me proud, you did it all."

The End

**Coal —— Tammy Lamb. Copyright,
<u>Tamlamb@adelphia.net</u>**

Ship :Immigrants leaving ship
Statue of Liberty
Passengers getting examined ——
Copyright, Aramark Sports and Entertainment,
Inc.
Aramark is an authorized concession of the Ellis
Island
Immigration Museum, National Park Service,
United
States Department of the Interior

Aramark Sports and Entertainment Inc, Ellis Is-
land
Immigration Museum, New York, NY. 1004
Phone: (212) 344-0996

Passengers disembarking ship
Ellis Island views
Immigrants leaving ship
Registry room —— Nick's World Immigration
Picture Page.
E-mail- <u>NDunklee@Mail.Davison.k12.Mi.US</u>

Excavator with wheel ——- Encyclopedia

Company Homes — Copyright, by: George Hvan
From the book: (When the Mines Closed)
**Publisher permission, Cornell University
Press,Sage House**
512 East State Street, Ithaca, New York. 14850

Miners Houses
Mine cave-in — Vintage Photographs From Yesteryear
                Copyright by: John Schehrer.
E-mail :
20jscherer@gbronline.com?cc=&bcc=&subject=&
body=

Miners bath
Explosion
Mining procedures
Miner and canary —— Staffordshire Multimedia
Archive

Working colliery — Copyright, Coal Region Enterprises.
E-mail :   • Jay Schutawie (webmaster)
webmaster@coalregion.com

Rubbish —— FreeFoto.com

Landfill —— copyright by: David Kovar

Stephen A. Wzorek

Comments and questions:
kovar@csst.org

Miner Wisdom

About the Author

Miner Wisdom is the original idea of myself, Stephen A Wzorek. I am 55 years old and a plant-engineering supervisor for a military defense contractor. Over the years I have been reading articles and seeing different news casts of devious attitudes of various business people. In the 1990's, in the area I live, there was blasting for a new highway near a major landfill. The adjoining development was sensing odors of gas. I believed they were looking in the wrong place for the problem. It was my belief that the odor was coming from the landfill, contributed by the disturbance of the mines. It was that thought which inspired the novel.